POCKET GUIDE TO Sewing Notions

Carry-Along Reference Guide

by Sue O'Very-Pruitt

POCKET GUIDE TO SEWING NOTIONS

Landauer Publishing, www.landauerpub.com, is an imprint of
Fox Chapel Publishing Company, Inc.

Project Team
Editorial Director: Kerry Bogert
Editor: Amy Deputato
Copy Editor: Colleen Dorsey
Designer: Mary Ann Kahn
Photographs by Sue O'Very-Pruitt and Heidi O'Very

ISBN 978-1-947163-43-0

Library of Congress Control Number: 2020037830

We are always looking for talented authors. To submit an idea,
please send a brief inquiry to acquisitions@foxchapelpublishing.com.

Printed in Singapore
23 22 21 20   2 4 6 8 10 9 7 5 3 1

Contents

What Is a Sewing Notion?

The word *notion* has been part of the English language for centuries and, over the years, has acquired a variety of definitions. In the late fourteenth century, it was used to mean "idea" or "concept." Over time, the word took on additional meanings, referring to how something should be done or a person's desire to do something. In more recent times, it's also been used in a snobbish tone, meaning "silly idea." By the nineteenth century, *notion* had also come to mean "small objects or tools used in sewing," and that is precisely how I will be using the word in this book. There are thousands of sewing notions out there, and being that this is a pocket guide, I will break them into general categories, sharing an overview of items and limited lists of specific products and brands.

ByAnnie.com

All About Sewing Notions

Specialty adhesives designed for use with fabric hold pieces together and are also used to position appliqué and trim.

ADHESIVES

An adhesive is a product with some sort of sticky substance that works to hold one thing to another—in most cases, trims and fabrics.

Glue: Fabric glue is by far the fastest and most non-invasive way of holding two materials together for sewing in a wide range of projects. Glue sticks and pens typically come in two sizes: ½" (1.3cm) diameter and ¼" (0.5cm) diameter. The wider sticks are ideal for larger projects, like adding trim to a purse or garment, while the narrower pens are ideal for smaller projects and working in tight spaces. Because the glue stick/pen works like a marker or regular pen, you can control the placement and amount of glue, which is especially helpful when working with small pieces of trim and appliqués. Refillable glue pens are slightly more expensive, but once you buy a pen, you only have to pay for refills.

- June Tailor®—Fabric Glue Stick: This non-toxic glue stick with a ½" (1.3cm) diameter holds fabric in place for quilting, appliqué, hemming, and general sewing. It goes on purple, dries clear, and washes out. It also works well for English paper piecing, basting, and positioning fine trim.
- Dritz®—Fabric Glue Stick: Use this glue for basting fabric, positioning trim and appliqué, matching plaids and stripes, and altering patterns. It is water soluble, acid free, and non-toxic, with a ½" (1.3cm) diameter.
- Sewline™—Water-Soluble Fabric Glue Pen: This water-soluble glue starts out blue and dries clear. Use this product to hold fabrics together or position pieces for sewing. This ¼" (0.5cm)–diameter glue pen comes with one refill, with additional refills available in packages of two.

Seam sealant: Even though seam sealant isn't actually an adhesive that holds two fabrics together, I placed it in this section because it has some similar characteristics. For example, it is a liquid that dries on the fabric. Seam sealant is used to stop a fabric or trim from fraying and to seal loose thread tails. It is used very often with satin ribbon, which starts to unravel almost immediately when cut. If you apply seam sealant to the end of the ribbon quickly after cutting, you will stop the fraying.

The texture of seam sealant once it's dry varies from product to product; it ranges from soft and pliable to slightly crisp. Let the seam sealant dry completely before touching it, because, depending on the fabric, the liquid can take on the color of the fabric and transfer it to another surface. I recommend always letting the seam sealant dry on a non-porous surface or a white sheet of paper. When the seam sealant has thoroughly dried, you can wash or dry-clean your fabrics.

- June Tailor®—Fray Block: With an ultra-fine precision tip, this product prevents fraying on fabric and ribbon. It dries quickly, is clear, soft, and flexible, and is suitable for washing or dry cleaning.
- Dritz®—Fray Check: This product keeps fabric from fraying and secures thread ends; it is also available with a fabric-guide applicator tip. It's suitable for washing or dry cleaning.

Spray adhesives: Spray adhesives allow you to cover large surfaces quickly with a light adhesive mist. Use a spray adhesive when putting two pieces of fabric together, for example, when making a placemat and applying the front fabric to the batting. If you are working on a large project, like a quilt, you should spray half of it at a time; otherwise, the glue from the spray will dry before you've had a chance to smooth out all the air bubbles. Look for a spray that is tacky while still allowing you to reposition the fabric. Be sure to pick a spray adhesive designed for sewing, because sprays designed for paper can leave spots on your fabric and gum up your needle. Some brands make adhesives that work with both fabric and paper.

- Odif—505® Temporary Adhesive: This temporary high-tack adhesive is odorless and colorless; it can be used for quilt basting, appliqué, and machine embroidery without a hoop. It leaves no residue and will not gum up needles.
- Sulky®—KK 2000™: This spray temporarily bonds fabrics together for appliqué as well as batting, other stabilizers, patterns, and templates. It allows you to reposition the fabric before stitching. This product adheres material to stabilizers to significantly reduce the movement of the material while embroidering. The formula will not gum up needles.
- Dritz®—Spray Adhesive: Use this temporary adhesive spray instead of basting and pinning. It is good for bonding fabric or paper, it allows repositioning, and it will not gum up needles.

Tape: There are two main types of tape for sewing, but they both serve the purpose of holding materials together temporarily for sewing. The first type is wash-away embroidery tape, which can be used for many applications other than embroidery—even for tasks like attaching a zipper to a skirt. You can sew through the tape without it gumming up your needle, and the tape will not get trapped in your project. It perforates upon being penetrated by a needle, and any small bits that remain will wash out.

The other type of tape is double-sided transparent tape, which also has many uses, from attaching trim to folding up hems. This is also a wash-away tape, but it comes on a paper backing that you must remove before stitching. For example, when creating a hem, first unroll the tape and stick the adhesive side to the fabric along the raw edge. Firmly finger-press the tape in place and then remove the paper backing to expose the other side of the adhesive. (Hint: Use a straight pin to pull the paper up and remove it easily.) Then fold up the hem as desired and finger-press to hold it in place. After you stitch the hem, wash the tape away.

- OESD™—Expert Embroidery Tape WashAway (¾ inches x 8 yards [2cm x 7.3m]): Use this wash-away tape to temporarily hold your fabrics in place for in-the-hoop embroidery and many other projects. Don't worry about getting tape trapped in your project, as this tape can be removed easily with water.
- Dritz®—Wash Away™ Wonder Tape: This double-sided, ¼" (0.5cm)–wide transparent tape is removed with just one washing. It secures seams and hems before stitching, and you can sew right through it without gumming up your needle.

BIAS AND BINDING

A binding is a piece of material that encases the raw edges of another piece of material to keep the fabric from fraying. Here, we discuss tools to help make binding (on the straight of grain and on the bias) and templates that help attach the binding.

Bias tape and binding makers: The difference between bias tape and binding is the way the fabric strip is cut. Bias tape is cut on a 45-degree angle, allowing the fabric to have the most stretch, whereas binding is cut on the straight of grain, which will have little to no stretch. Bias tape is typically used for covering raw edges around curves, like on a baby bib. Binding is usually attached to straight-of-grain projects like placemats or quilts.

When cutting the strips on the bias or straight of grain, cut the width based on the tool you are using and follow the instructions. The most popular cutting width for bias is 2" (5cm), which is then double-folded to its final width of ½" (1.3cm). The most common width for binding is 2½" (6.4cm), then folded in half. Folding these strips manually can be tedious and inaccurate. Using a bias tape or binding maker takes the guesswork out, as you simply cut the strips to the instructed width and feed them into the maker.

- Clover®—Bias Tape Maker: Use this with a strip of fabric (the width will vary depending on the desired final width). Cut one end on a 45-degree angle and feed it into the larger open end, then pull it out on the other end as you are ironing. This will force the raw edges to meet, and you will press them flat. When you are finished ironing the full length, align the two folded edges together and press one more time manually. This works with strips cut on the bias or straight of grain. The following sizes are available, based on finished width: ¼" (0.5cm), ½" (1.3cm), ¾" (1.9cm), 1" (2.5cm), and 2" (5cm).
- June Tailor®—Quilt Binding Express: This tiny pressing mat is designed to press binding strips quickly and easily, much like a bias tape maker. The three strip sizes are: 2" (5cm), 2¼" (5.7cm), and 2½" (6.4cm). Simply feed the strip through the slot to fold and press right on the mat.
- Purple Hobbies—2" (5cm) Double Fold Binding Tool for ½" (1.3cm) Strips: Like the name indicates, this bias tape maker does it all in one pressing action. Cut a strip 2" (5cm) wide and cut one end on an angle to feed it into the wide opening more easily. You can press it flat and open, or you can flip the tool on its side and press both folds at once. It comes in a variety of colors.
- Purple Hobbies—The Third Hand Binding Folder Clip: This notion is adjustable, so it fits standard ironing boards as well as thinner boards. Loosen the wing nut, slide the Third Hand into position, and turn the wing nut to secure it in place. To create binding, insert a 2½" (6.4cm) (or, if you prefer, 2¼" [5.7cm]) strip of fabric into the clip. The slot on top allows you to use a pin to help pull the fabric through as needed. Pull the binding onto your ironing board and press as you pull. This item comes in a variety of colors.

Binding accessories: Once you have your strips of fabric, it is helpful to have a holder that keeps them flat and feeds them off evenly—especially when it comes to quilting binding, which can be several yards (or meters) long. The notions in this section are cute, but, more importantly, they are designed to perform the function of holding and delivering the bias or binding in a uniform manner.

Bias tape makers and binding tools finish raw edges for a clean, professional look. (Fabric from Christa Quilts.)

- Doohikey Designs®—Binding Babies®: These sweet little notions are both adorable and functional. These "babies" are made of wood and painted in cute colors. The top is the head, and the body looks like a spool with a slot, which holds the binding. Simply slip your binding, bias, or trim into the slot, wrap it around the body, and hold it closed with a pin or clip. They come in two sizes: medium measures 4" x 2" (10 x 5cm) and holds 2" to 2½" (5 to 6.4cm)–wide trims, and large measures 6" x 2" (15.2 x 5cm) and holds up to 3¾" (9.5cm)–wide trims. There are also limited-edition and seasonal Binding Babies as well as an option to design your own.
- Doohikey Designs®—Flower Spindle™: This pretty, flat wooden flower is painted, has non-slip pads on the bottom to keep it from sliding on your sewing table, and has a spindle in the middle. The spindle can hold one of your Binding Babies, enabling it to hold more binding by providing a resting place for it. It measures 1" (2.5cm) tall by 6" (15.2cm) wide.
- Purple Hobbies—Binding Wheel/Binding Wheel Clamp: The binding wheel resembles a small Ferris wheel and helps you organize your folded binding by rolling it around the wheel. The binding will then unroll smoothly as you're sewing it to your project. The clamp, available

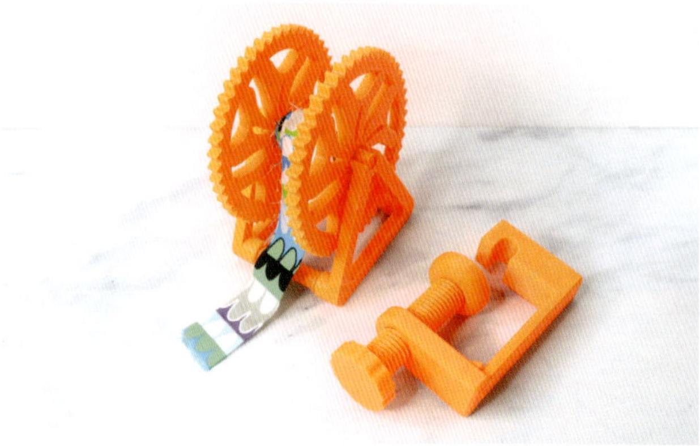

The Binding Wheel by Purple Hobbies is a clever invention designed to hold and smoothly feed your bias tape or quilt binding. (Fabric from Christa Quilts.)

separately, secures the wheel to your work surface or ironing board, allowing you to press and roll onto the wheel with a smooth motion. This notion comes in a variety of colors.

Binding templates: After you have sewn the binding to a project—for example, around the outside of a quilt—you will need to attach the two ends. You do not want bulk inside the seam; you want the binding to lie flat on the raw edge of the quilt. A binding template simplifies this process by helping you cut at the indicated spot on the binding and then stitch, fold, press, and finally sew the binding end onto the quilt.

- Quilter's Paradise—CutRite™ Bind Up: Easily finish a quilt binding with this handy, no-measuring-required tool. The fluorescent green color is easy to see on all fabrics. Just fold in the binding with 5" (12.7cm) to spare, mark, and stitch.
- TQM Products—The Binding Tool: This tool helps you angle the end of your binding when finishing a project to reduce the bulk of this last step. Use it to make two measurements for two cuts. This helpful notion comes in two sizes: the Binding Tool is for larger projects, and the Mini Binding Tool is for smaller projects, like potholders.

CUTTING TOOLS AND MATS

To sew a finished project, you must first cut the fabric, pattern, and thread. This section discusses various cutting tools and cutting surfaces.

Rotary Cutters

Rotary cutters are like pizza cutters. When you use a pizza cutter, you want to cut on a protective surface so that you don't ruin your countertop. The same principle holds true for the rotary cutter. Rotary cutting has a few required pieces: the rotary cutter, the rotary cutting ruler, and the rotary cutting mat.

There are many different brands of rotary cutters and accessories. To demonstrate the difference between an ergonomic and a straight cutter, I'll use the Olfa® brand as a point of reference.

The Ergonomic Rotary Cutter has an angled handle. It is easy to hold, easy to open and close, and comfortable when cutting multiple layers. This type of cutter has a self-retracting blade with a locking button to keep the blade either exposed or hidden. Designed to work for both left-handed and right-handed users, it comes in two sizes. The smaller size, 45mm, comes in different colors and can work with specialty blades. The larger size, 60mm, comes in one color and works with only the standard blade.

Rotary cutters come with straight or angled handles, and there are specialty rotary blades designed to create decorative edges.

The Straight Handle Rotary Cutter has a straight handle, but it is contoured for comfortable cutting. The blade must be opened and closed manually. It is also designed for both left-handed and right-handed users, and it comes in four sizes: 18mm, 28mm, 45mm, and 60mm.

Specialty blades and accessories: Because you are working with sharp objects, safety is important, and that includes using the correct blades. Whichever brand of rotary cutter you decide to use, I recommend using blades of the same brand. Double-check that the blade you would like to use works with your rotary cutter. Just because the blade is the correct size does not necessarily mean that the rotary cutter is designed to work with it. If you use the wrong type of blade, you could damage the blade, the cutter, or the cutting mat.

- Olfa®—Endurance™ Blade: This expertly honed blade offers long-lasting cutting performance and durability and is available in 45mm and 60mm sizes.
- Olfa®—Rotary Pinking Blade: Use this blade with the Olfa® 45mm Ergonomic Rotary Cutter to provide a zigzag pattern effect. This blade has "teeth" on the edge to give the fabric a decorative and functional finish.
- Olfa®—Scallop/Peak Rotary Blade: This blade is also used with the Olfa® 45mm Ergonomic Rotary Cutter. It's two blades in one: one side creates a scallop design, and the other creates a peak design.
- Olfa®—Wave Rotary Blade: For use with the Olfa® 45mm Ergonomic Rotary Cutter, this blade has a soft wave edge and is popular for crafts.
- The Gypsy Quilter®—Rotary Blade Guards: These are clear plastic circular guards; apply one under your 45mm rotary blade for added safety while cutting.

More Cutting Tools

Seam rippers: Seam rippers often have a negative connotation because people associate them with making mistakes. The truth is, though, that they come in handy for a lot more than just ripping out stitches. Some can be used to grab and hold a single thread, and others can cut a buttonhole with ease. It is important to know that seam rippers typically have a life span of about six months to a year, depending on use, so you may want to consider a seam ripper with replaceable blades.

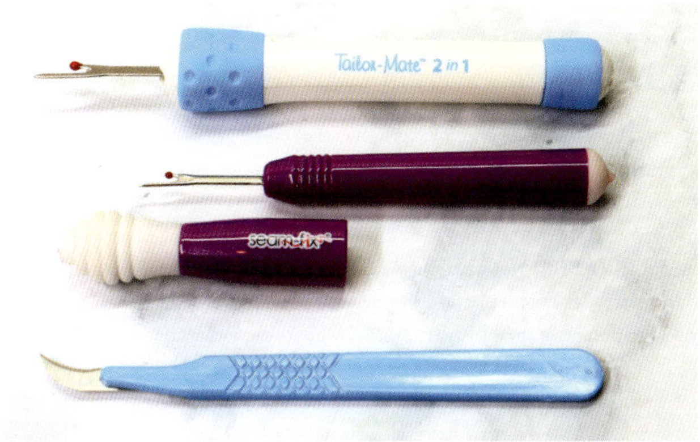

Seam rippers come in many shapes and sizes, with sharp blades to make precise cuts.

- Famore Cutlery—Easy Glide Surgical Seam Ripper: Featuring a "power-lock" screw design, this seam ripper's blade is considered "surgically sharp." The curved blade glides and removes serger stitches with ease. Replacement blades are available.
- Sullivans—Precision Seam Ripper and Buttonhole Cutter: This small, thin, stainless-steel blade is a very helpful tool for the home sewer. Made with surgical quality, this seam ripper easily digs in and takes out stitches and seams. Also included is a buttonhole cutter to quickly create a perfect ⅜" (1cm) buttonhole. This seam ripper comes with a protective safety case and an extra blade and is available in several colors.
- Dritz®—Seam-Fix™ Seam Ripper: In addition to the traditional seam ripper with a point, curved blade, and red ball on the tip, this tool has a special plastic material on the ends of the handle and cap that will grab and remove excess threads once you've ripped out the seam.
- Taylor Seville®—Tailor-Mate™ 2-Way Retractable Seam Ripper: This two-sided seam ripper offers retractable large and small seam rippers. The ergonomic handle offers comfort and precision, and the rubber ends help remove cut thread.

Scissors: A good pair of scissors can change your cutting experience. Not only is the quality of the blade important, but choosing the right type of scissors allows you to be more efficient and accurate. The reason there are so many brands and types of scissors is that there are so many different applications, projects, and types of materials. In this section, we are going to discuss products made by Famore Cutlery unless otherwise noted.

Shears

- EZ Glide Fabric Shears: These 11½" (29.2cm) shears are designed to cut batting easily and not snag while cutting a range of fabrics, including silk, polyester, wool, and more. They are designed with bent rubber handles for comfort and balance for tabletop cutting.
- Pinking Shears: Pinking shears, also known as scalloping shears, cut the fabric in a sawtooth zigzag pattern. They are used mainly to minimize fraying and reduce bulk in seams.
- Pro Cut Comfort Handle Fabric Shears: These 8" (20cm) shears with cushioned handles are designed for sewers with arthritis and for long-term use. They are micro-serrated for steady, controlled cuts.

Appliqué scissors

- Duckbill Appliqué Scissors: These scissors offer precision for trimming around the edges of appliqué, trimming stabilizers from the back of your embroidery, and cutting close to seams without cutting the back

Not all scissors do the same thing. Shown (left to right) are Famore Cutlery's EZ Glide Fabric Shears, Pinking Shears, and Pro Cut Comfort Handle Fabric Shears.

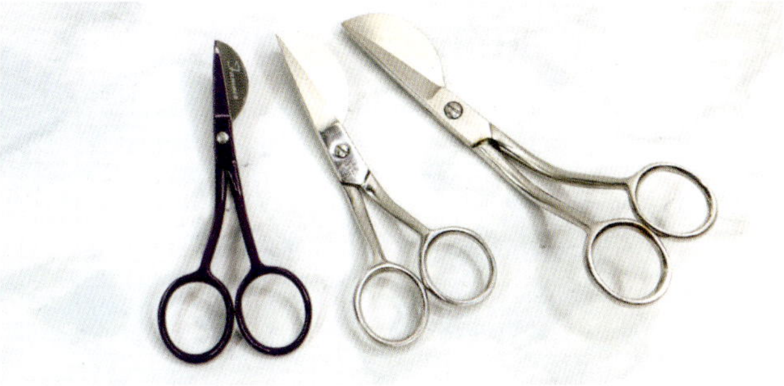

Duckbill appliqué scissors have a large curved bill on one side, allowing for accurate, close cutting.

side. The duckbill lies against the main material under the appliqué fabric, while the straight blade is inside the appliqué area. Comes in 4" (10cm), 4½" (11.4cm), and 6" (15.2cm) sizes.

Thread snips

- EZ Stitch Snip with Hoop Blade: These come in two sizes—4½" (11.4cm) and 6" (15.2cm)—and have a hook that can catch and hold the thread and then snip. These snips allow both right-handed and left-handed use.
- EZ Stitch Snip with Micro-Serrated Blade: These are good for snipping stitches and cutting tail ends. The curved end makes getting into tight areas easy, and the fine tip allows you to get under tight stitches. These allow both right-handed and left-handed use and come in either silver or with rainbow coating.

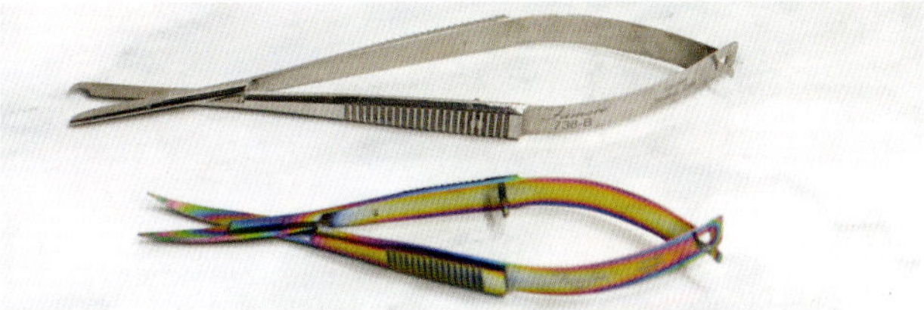

One of the functions of thread snips is to get in between stitches to clip the thread.

LEFT-HANDED SCISSORS

The difference between right-handed and left-handed scissors is the outside blade. With left-handed scissors, when you hold the scissors in your left hand, the outside blade will lift toward the top. The thumbhole is what controls the lower blade and the direction of the scissors. Famore Cutlery has an entire line of left-handed options, and here are just a few:

- **Left-Handed Duckbill Appliqué Scissors: These come in 4½" (11.4cm) and 6" (15.2mm) sizes and are ideal for cutting appliqué.**
- **Left-Handed Large Ring Fine Tip: These scissors feature a sharp tip with large rings for all hand sizes.**
- **Left-Handed Tabletop Fabric Shears: These are basic left-handed fabric shears.**

Thread cutters: There are many ways to cut threads, and there are many reasons that threads need to be cut. These thread cutters are designed to help cut threads between sewn fabric pieces. Must like assembly-line sewing, chain piecing is the process of sewing several pieces in one continuous seam and then cutting the threads between each piece. You could use small scissors or thread snips, but using a thread cutter is the most efficient option.

- Purple Hobbies—Bladesaver Thread Cutter: This clever invention takes an old rotary cutter blade and turns it into a side cutter that's good for cutting chain-stitched quilt blocks. Unscrew the flower pieces and place a used rotary blade inside, then replace the flower and screw. The carrying case doubles as a stand with small gripper feet at the bottom, so you can set the flower inside the stand and begin cutting. This product is available in a variety of colors.

The Bladesaver allows you to recycle an old rotary cutting blade and use it to cut chain threads.

- Dritz®—Thread Cutter Caddy: This is ideal for cutting threads between chain-pieced fabric. Small items such as pins or thimbles fit handily inside the caddy, and the sides have magnets to hold pins and needles.

Mats

Using a rotary cutter can make cutting quicker and more efficient, but you will need the proper surface to cut upon—and that's where cutting mats come in. A rotary cutter can damage your work surface, but a cutting mat will protect your surface and provide measurements, lines, and angles for accurate cutting.

Cutting systems

- DIYStyle®—Magnetic Pattern and Cutting System: More than just a mat, this is a three-part system that eliminates pattern weights and pins. The 24" x 36" (61 x 91cm) multi-use, self-healing mat is white with easy-to-read black markings on one side. On the other side is a pattern-marking cork surface for creating or adjusting patterns. It comes with a 36" (91cm) T-square and alignment tool that is specially designed to work with the mat. There are holes in the T-square that hold the

The DIYStyle cutting system is designed to make measuring and cutting more accurate and efficient. (Fabric from Christa Quilts.)

included FabricMagnets™, which can adhere to both sides, making cutting and designing patterns easy.

Rotary cutting mats

- Olfa®—Double-Sided: Designed to be used with Olfa® rotary cutters, this mat has a self-healing finish to provide a continually smooth surface. The double-sided mats are 1.5mm thick, are constructed with a triple-layer heat-welding process for a long life, and come in a variety of sizes and colors.
- Olfa®—Rotating: This is a non-slip mat comprising a stationary bottom layer and an interlocking top cutting mat that rotates 360 degrees. The self-healing finish provides a continually smooth surface. It comes in green in two sizes: 12" (30.1cm) and 17" (43.1cm).

Rotary cutting mats come in different shapes, sizes, and colors to protect your work surface.

HAND-SEWING TOOLS

Hand sewing requires a few different notions than machine sewing, and once you have them in your sewing room, you will quickly realize that you cannot live without them.

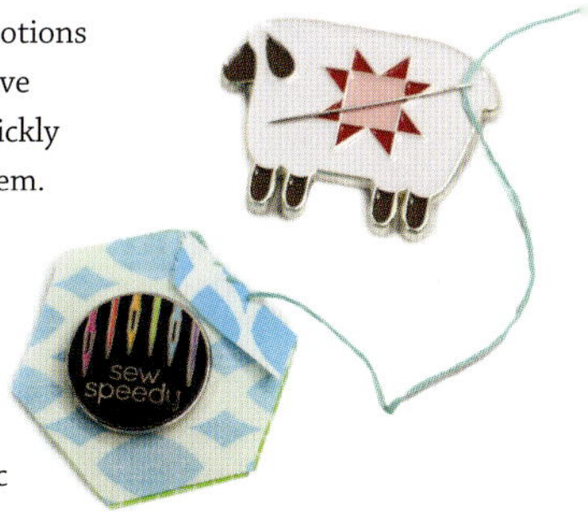

Needle minders are often fashionable in addition to being functional. (Fabric from Christa Quilts.)

Needle minders: Sometimes, when hand sewing, you put your needle down and lose track of it until it pokes your finger. A needle minder takes up very little room on your sewing table and has a strong magnetic surface to hold needles in place so you can find them every time. You can clip some needle minders to your project or clothes to keep your needles close at hand.

- It's Sew Emma®—Prim Sheep Enamel Needle Minder by Lori Holt of Bee in My Bonnet Co.: This adorable needle minder is designed to coordinate with the Prim Quilt Kit and Prim Stitch Series patterns.
- Whole Circle Studio—Sew Speedy Enamel Needle Minder: This colorful 1" (2.5cm)–diameter enamel needle minder is designed to hold your pin or hand-sewing needle in place for a moment. The back side has a second, detachable magnet, so you can attach the needle minder to your shirt, your sleeve, or even the project you're working on and always know where your needle is. You can even use it as a fashion accessory!

Thimbles: Hand sewing means that you are manually pushing and pulling your needle through the fabric, and this can wreak havoc on your fingertips. A thimble is a small cap that fits over your fingertip to protect your skin while sewing.

- Clover®—Protect & Grip Thimble: This soft elastic finger protector has a scallop design for breathability and a ridged and dimpled metal thimble cap for safety. It provides comfort and lightness for use in a wide range of sewing applications and comes in three sizes: small, medium, and large.

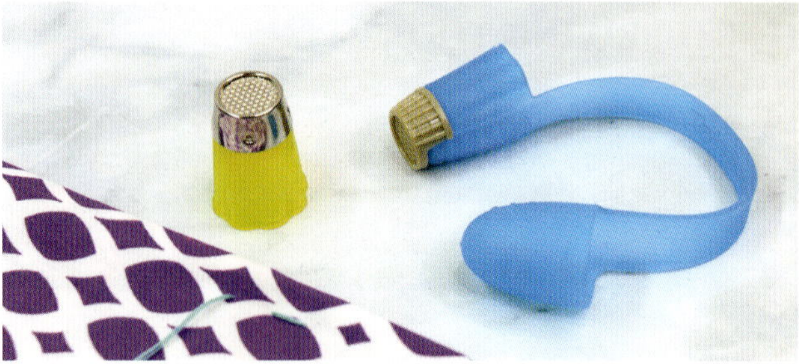

Thimbles are popular notions because they help you push, pull, and hand sew without hurting your fingertips. (Fabric from Christa Quilts.)

- Taylor Seville®—Wonder Grip & Thimble™: Place the nubby part on your thumb and the actual thimble part on the finger of your choice. This finger protector is soft and flexible, grips needles firmly, and is easy to push in and pull through. On the thimble part, there are dimples and ridges that help keep the needle from slipping.

Threaders: Getting your thread through the eye of a needle can sometimes be so frustrating that it discourages you from sewing! To make the process easier, use a needle threader, which is designed to help you thread needles with minimal effort.

Putting a thin piece of thread through the tiny eye of a needle can be frustrating, if not impossible, without the help of a needle threader.

- Clover®—Desk Needle Threader: This handy notion threads your hand-sewing and quilting needles quickly. Making sure that the thread thickness is suitable for your needle, insert the thread and hold it gently, then insert the needle eye down into the needle hole, press the lever once, pull the needle out, and cut the thread. The threader comes in a variety of colors.

- Dritz®—Hummingbird Needle Threader: This wire threader has an easy-to-hold shape and an indented profile for finger placement and gripping. The beak swings upward to expose the wire and downward to protect it for storage. Small holes on the body and wing can hold lanyards. It is suitable for most hand needles except beading needles.

Thread conditioner:

The main reasons to use thread conditioner are to strengthen the thread, prevent it from fraying, and help reduce fatigue when sewing. Some brands provide added benefits, such as protecting thread from UV rays, preventing knotting and tangling, and extending the life of finished projects.

Thread conditioner helps strengthen your thread throughout the life of the project.

- Wise Craft—Sew Fine Thread Gloss: This beeswax-based thread conditioner coats your thread and prevents knotting and tangling. Use it by threading your needle and laying the thread close to the needle on the thread gloss. Place light pressure on the thread with your thumb while you pull the thread along the top of the thread gloss. One or two light pulls should be plenty, and then you can knot your thread and start stitching. Available in a wide range of scents, including cotton candy, strawberry, and pineapple, as well as exclusive scents from Wise Craft such as Earl Grey, Hygee, and Rainfall.
- Taylor Seville®—Thread Magic®: This product is designed to make thread more manageable and extend the lifetime of finished work for many years. It comes in a cube with a sticky backing for attaching to the front of your sewing machine. Press the thread into the gel, thread the

needle and adjust the tensions, and run the machine. It's also sold in a round two-part case: the top part is the lid, and the grooved base holds the conditioner; the grooves along the side make it easier to slide the thread through the conditioner. The acid-free conditioner eliminates static, does not melt or freeze, and is safe for all threads and fabrics.

Thread minders: Just as a painter has a palette to lay out paints, a hand embroiderer likes to have the various threads for a project separated and organized, which can be critical when following a pattern. A thread minder is a tool that makes this possible, and some brands offer bonus features such as thread guide cards and a lens for viewing color values.

- Wise Craft Handmade—Ruby Minder™: This thread minder keeps your embroidery thread organized and doubles as a design lens tool to view color values in your needlework. It holds up to twelve embroidery thread colors. The designer even suggests wearing the minder around your neck as a pendant to sew on the go.
- It's Sew Emma®—The Beekeeper Thread Minder and Thread Guide Set by Lori Holt of Bee in My Bonnet Co.: Keep your hand embroidery and cross-stitch projects organized with this thread minder. For projects with multiple hues, keep each color tied to a different numbered slot, then record the color names with the corresponding number on the thread guide as a reference.

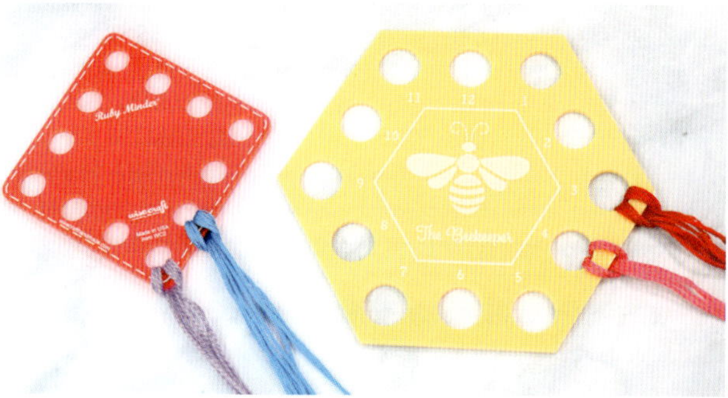

Thread Minders keep your threads and embroidery floss separated and orderly.

IRONS AND PRESSING TOOLS

Pressing a project with the proper tools can take it from looking homemade to looking professionally constructed in just moments. From the iron to the board, these notions are designed to help you get crisp seams with professional results.

Irons: The purpose of an iron is to smooth out wrinkles, adhere fusible webs to materials, and more. When selecting an iron for your sewing space, consider factors such as price, durability, steam, and cord length (or cordless). An iron is an amazing example of science coming to life as it achieves the correct temperature for the best results. So, as simple as it sounds, read your iron's user manual to get the most out of it.

Travel-size irons are little but mighty! Choosing a smaller iron depends on the project. Do you require steam, a small pressing footprint, or even heat in a tiny space? Do you want an iron that's easy to bring to quilting groups or sewing classes?

Full size

- Panasonic®—360 Cordless Advanced Ceramic Plate: Without having to worry about a tangled power cord, you can enjoy 360-degree ironing with this steam/dry iron that comes with a charging base and carrying case. The dual-tipped stainless-steel soleplate allows smooth,

It's important to find an iron with the right features for your sewing space and the types of projects you create.

multi-directional ironing. You can adjust the concentration of steam as you iron a variety of fabrics. For safety, the charging base deactivates after ten minutes if left plugged in and unused. The lightweight carrying case makes this cordless iron easy to bring along to quilters' classes and sewing or crafting groups.

- Oliso®—PRO™ TG1600 Smart Iron: Touch the handle, and the iron lowers, ready to work. Release your hand, and the patented Scorchguards lift the iron off the board to prevent burns and tipping. Benefits of this iron include 1800 watts of power for faster ironing time as well as horizontal or vertical bursts of steam. Additional features include a thirty-minute extended auto-shutoff, large side fill tank, stainless-steel soleplate, flat pressing surface, detailer tip, and 12-foot (3.7m) cord for a full range of motion.

Travel size

- Cricut®—EasyPress Mini™: Measuring approximately 2" (5cm) wide by 3⅓" (8.4cm) tall, this handy press is a good option for small, portable pressing. Offering dual heating elements and a ceramic-coated plate, it delivers dry, even heat all the way to the edge to provide consistent results. It has three temperature settings and comes with an insulated safety base.

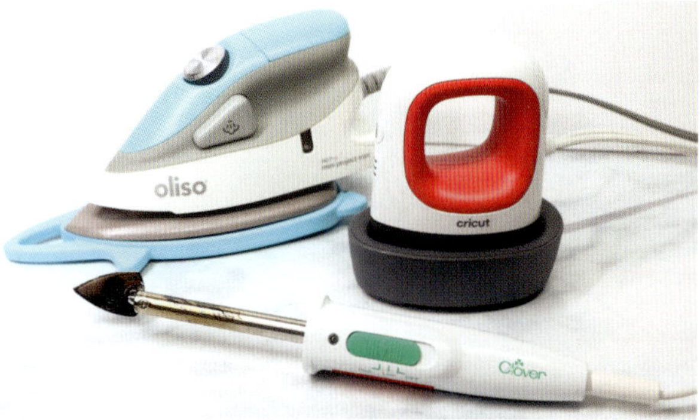

A mini iron is a helpful tool when working on small projects or attending sewing classes.

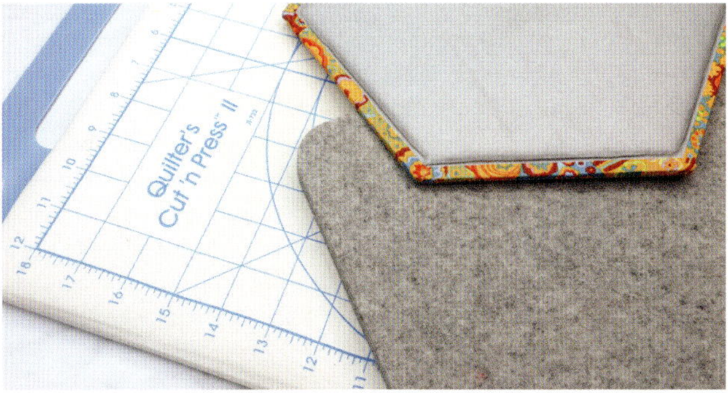

A pressing mat or pad can often take the place of a traditional ironing board.

- Clover®—Mini Iron: This mini iron works well for pressing quick bias in stained glass quilting, appliquéing, paper crafting, and many other quilting and sewing projects. It offers low and high temperature settings and an easy-to-grip handle. With its compact size, it's ideal for detailed work. A stand is included with the iron.
- Oliso®—The Mini Project Iron™: This iron has the power of a full-size iron in a compact design. It comes with the dual-purpose Solemate™ silicone trivet that hugs the mini iron for travel and storage and functions as a safe place for the iron to rest while in use. It offers a fabric selector, steam control, 360-degree swivel cord, and precision tip, and it comes in pink, yellow, blue, and purple.

Ironing boards/pads: Just like you choose your iron for the best results, the same goes for your pressing surface. Make sure that your ironing board, pad, or mat matches your iron in size; for example, when working on a small mat, use a small travel iron. This will help concentrate the heat and speed up the pressing process.

- June Tailor®—Cushioned Quilter's Square 'n Blocker: This portable pressing, squaring, and blocking board is a convenient 18" x 24" (45.7 x 61cm) size with a printed 14" x 20" (35.6 x 50.8cm) grid. A thick, bonded, cushioned layer eliminates seam-edge impressions and protects the raised details of fine appliqué. A durable, heavyweight inner board keeps the pressing surface flat and smooth, allowing you to press and square the smallest quilting pieces with complete accuracy. It's also handy for blocking needlework and yarn projects.

- June Tailor®—Cut 'n Press™: With both cutting and pressing surfaces and a handle for carrying, this is a great option for bringing to classes or working in a small space. The hard-surface cutting mat on one side offers durability and extended blade life, while the cushioned cotton/poly pressing surface on the other side is bonded to minimize stretching and distortion. It comes in 11" (27.9cm) square and 12" x 18" (30.5 x 45.7cm).
- Traditional Primitives—Hexie Pressing Mat: This is a portable pressing mat, made for use with the Clover® Mini Iron, that has a thermal ironing surface with a cute, colorful border trim unique to each mat.
- The Gypsy Quilter®—Wool Pressing Mat: This ½" (1.3cm)–thick, 100% felted wool pressing mat helps you press any pieced or embroidered block with ease. The mat's density absorbs seams and stitches, effectively pressing both sides at once. You can block knitted projects as well by pinning them directly to the mat. The texture of the wool stops fabrics from shifting so there is no distortion when pressing. You can use the pressing mat with a dry iron on any surface. If using steam, place the mat on a protective surface, because moisture goes through the mat and can harm the surface below. It comes in a variety of sizes.

Iron cleaners: As careful as you try to be, sometimes fusible webs and interfacings get stuck on your iron's soleplate. The following products will help make removing this buildup easier.

- Faultless®—Hot Iron Cleaner: This cleaner is designed to remove tough buildup of melted fusible glues and interfacings from the bottom of your iron in seconds, and it cleans hot glue guns too.
- Bo-Nash—Iron Clean: Each package contains ten reusable cleaning cloths that remove fusible interfacings and sticky residues from your iron plate. Just wipe your hot iron over the cleaning cloth with a paper towel or rag underneath.

- Dritz®—Iron-Off Hot Iron Cleaner: This non-flammable, non-toxic, and non-abrasive cleaner removes built-up starch, fusible webbing, and detergent from your iron.

Iron rests: Between pressing, give your hand and ironing board a break by setting your hot iron on an iron rest. This is especially helpful when using a small pressing surface so you can place your iron off to the side. Check whether it's OK to keep your iron in a horizontal position for extended periods and whether you should lower the heat when the iron is not in use.

- It's Sew Emma®—Sew Hot Iron Rest: This heavy-duty silicone iron rest in a minty hue allows you to place your iron plate directly on the surface, so you can "press on" without it tipping over. The generous 7" x 10⅞" (17.7 x 27.6cm) size will fit most home irons, and it can withstand up to 500°F (260°C).
- The Gypsy Quilter®—Silicone Iron Rest: Made with silicone, this iron rest resists heat, and the raised round dots also help the iron cool down faster. It measures 5⅜" x 10½" (13.7 x 26.7cm) and comes in red, green, or purple.
- Oliso®—TG Solemate: Fits the Pro™ TG1600 and several other Oliso irons. Designed to protect your iron's soleplate when not in use, the Solemate fits securely to the iron's soleplate, wrapping around the bottom corners and locking at the tip; it also features a built-in hanging

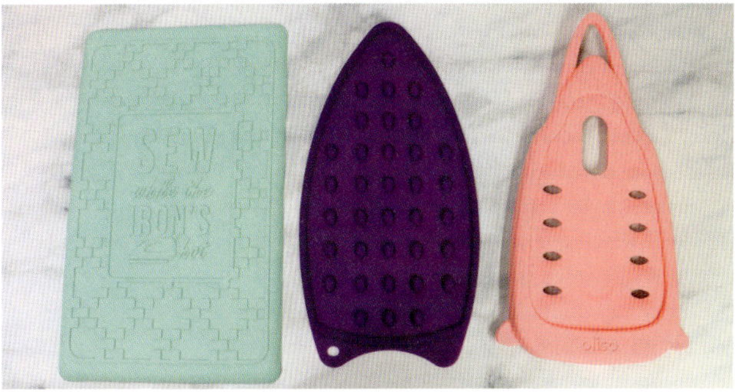

Use an iron rest as a safety measure and to protect your work and pressing surfaces.

Fusible webs can stick to your iron and pressing surface but will easily peel off a non-stick pressing sheet.

loop. When using the iron, flip the Solemate over, and it becomes an iron rest with steam holes for faster cooling.

Non-stick pressing sheets: It's a good idea to place a non-stick pressing sheet, rather than a piece of fabric, between your iron and fusible web. The difference is that any fusible residue left on the non-stick sheet will peel off easily.

- Bear Thread Designs, Inc.—The Appliqué Pressing Sheet and Multipurpose Craft Sheet: This double-sided appliqué pressing sheet is made from an ultra-high-temperature fabric and coated with non-stick Polylon. It comes in multiple sizes and includes a free appliqué pattern.
- Attached Inc.—Goddess Sheet®: With a non-stick quality thanks to Teflon®-coated fiberglass, this transparent sheet serves the dual purpose of reducing ironing time and putting a fine gloss on fused surfaces to let you know which side is which. Sometimes it can be hard to tell which side of your fabric has fusible web on it, so this helpful sheet allows you to fuse with confidence.
- The Gypsy Quilter®—The Silicone Appli-Fuse Mat: Made of a silicone and glass fiber material, this mat is heat resistant up to 500°F (260°C), making it the perfect surface for fusing, ironing, and hot-gluing. Because the mat is transparent, you can accurately place your cut

appliqué shapes, and, because it is heat resistant, you can place and press multiple layers simultaneously. It is slightly tacky, so the pieces won't migrate when positioned. This mat is easy to clean and available in multiple sizes.

Non-heat pressing: Some materials cannot be pressed with a traditional iron, even at a low heat, so it's nice to know that there are non-heat pressing alternatives. There are plastic and wood varieties, and the non-heat presser you choose will depend on the size of the seam and its location. Smaller non-heat pressers are ideal when working at your sewing machine because you can press the seam on the bed of the machine. Some of these tools work in conjunction with a heat iron, helping set the seam more firmly.

Plastic

- Clover®—Finger Press: This white plastic notion is designed to press small seams flat without having to use an iron. Shaped for easy use, it's perfect for foundation piecing, patchwork, and appliqué.
- Clover®—Roll & Press: This simple and clever notion won't pull or distort fabric, and it has an ergonomic handle that makes it comfortable to hold. The wheel is tapered to focus pressure on the

Consider non-heat pressing options for materials that cannot withstand the heat of an iron.

seam, preventing impressions on the right side of the project. This tool is ideal for laminates and sealing glued seams on leather and felt.

Wood

- ByAnnie—ByAnnie's Stiletto & Pressing Tool: The stiletto portion of this dual-purpose notion is covered in the Pointing, Turning, and Pushing Tools section. The pressing portion is at the perfect angle to create a surface for pressing seams open, so you don't need to move to the ironing board for just a small seam—you can press right at your machine. It's very lightweight and easy to hold, and it's curved on the side so it won't roll off your work surface. The handle is durable and water-resistant.
- Angela Wolf Patterns—Tailor's Clapper: This is a multi-use tool for every sewing room. As seen on It's Sew Easy TV, Angela Wolf uses this tool for everything from pressing up hems, creasing pants, pressing seam allowances open, and even pressing knit fabrics—all you need is a little steam. It comes in regular and mini sizes.

Press cloth: For some materials, even at low heat, a press cloth is recommended between the iron and the material.

- June Tailor®—Eze-View Pressing Cloth: This universal, 100% cotton press cloth protects and prevents shine on fabrics and allows you to see as you press. You can use it with steam or a dry iron; for extra moisture, dampen the cloth with a mist sprayer or sponge. It comes in a generous 14" x 24" (35.6 x 61cm) size.

Spray starch and alternatives: When fabric comes out of the dryer and is wrinkled, you can use a liquid starch or starch alternative to help remove those wrinkles. Some traditional spray starches can leave residue on your iron soleplate, leave spots on your fabric, or yellow your fabric over time, in which case using a starch alternative is a better option for the life of the fabric—especially with quilts that will be passed down from generation to generation.

- Soak®—Flatter: This starch-free smoothing spray relaxes wrinkles and freshens fabrics to make ironing easier. The soft mist sprayer provides even distribution with no flaking, clogging, or residue. Use it with steam or a dry iron on the recommended fabric settings.

The biodegradable, phosphate-free, and dye-free formula is made with plant-derived, renewable ingredients. It is available in various fragrances as well as unscented.

- Mary Ellen—Mary Ellen's Best Press: This spray starch alternative helps fabric resist wrinkles to make ironing easier, and the special stain shield protects fabrics. The acid-free formula leaves no flaking, clogging, or white residue on dark fabrics. It comes in a variety of scents as well as unscented.
- June Tailor®—Quilter's Starch Savvy: Formulated specifically for quilting cottons, this product also works on many other types of materials. It stabilizes fabric to avoid block distortion and create flattering seams, and you can control the level of stiffness with multiple applications. It also works to restore body to laundered fabrics. Fragrance- and flake-free, it comes in a refillable non-aerosol fine-mist spray bottle.

Spray starch alternatives help you achieve crisp seams.

Using a cutting mat on the light box simplifies the paper-piecing process. (Fabric from Christa Quilts.)

LIGHTING AND MAGNIFIERS

We want accuracy when sewing, and having enough light or magnification can make or break a sewing project.

Light box: There are a few things to consider when choosing a light box: brightness, size, weight, and portability. It's also nice if you can use a cutting mat on the light box for paper piecing; that way, you can cut right over the light on a protective surface.

- Daylight Company—Wafer: Ultra-slim (about ⅓" [8mm] thick) and ultra-light (2 lb [0.9kg]), with an extra-bright illuminated area, this is a must-have tool for tracing on paper or fabric. The LED light box provides even lighting from edge to edge, and the brightness is adjustable to suit the thickness of the paper or fabric. The handy marking guide along the sides gives measurements in inches and centimeters. A cutting mat that fits on top is available separately.

Magnifiers: When you need extra magnification for hand sewing or machine work, wearing glasses might not provide enough amplification to see the project details. You may need a magnifying tool to help you see your stitches up close.

- Taylor Seville®—CloseLook™: This lighted magnifier has an adjustable visor that swings up and down and can be locked in place. You can adjust the focus by moving the object closer to or farther away from the lenses. Three interchangeable optical-grade lenses are included.

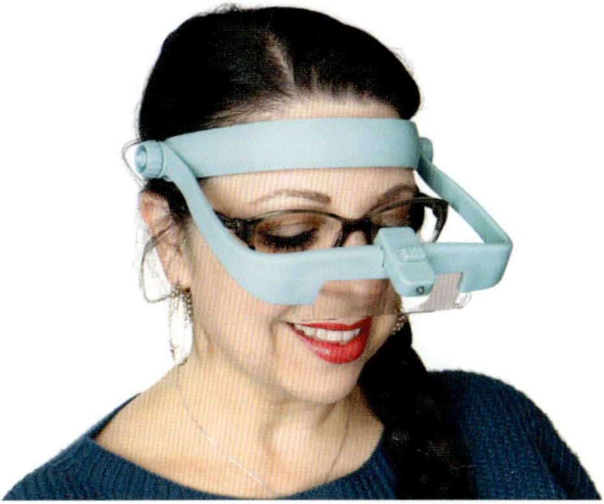

A magnifier does the trick when your glasses don't give you the up-close magnification you need.

- Taylor Seville®—Magic Eyes™: These clip-on eyeglass magnifiers with super-sharp quality allows you to see up to 3X magnification. A flexible arm allows you to make adjustments for reading, small hand-sewing work, and more.
- Daylight Company—MAGnificent Floor & Table LED Magnifying Lamp: Useful for intricate work, this two-in-one magnifying lamp can be quickly adjusted to a 50" (1.3m) floor lamp or a 20" (51cm) table lamp to fit the task at hand. The easy-to-assemble LED light provides true color matching, so you can carry out detailed work at any time of day or night. Get a closer view of your work with the large (almost 6" [15cm]) semi-rimless lens with 1.75X magnification. Use the flexible arm to direct light exactly where you need it. This tool is ideal for using with small hand-sewing projects, or you can position the lens right in front of your sewing machine needle.

Portable lighting: Portable sewing tools, including lights, are optimal for moving around your sewing room, taking to classes, or when you don't have a permanent sewing location. Some portable lights can even be attached to your project. The idea is that, with portability, you can keep sewing wherever you are.

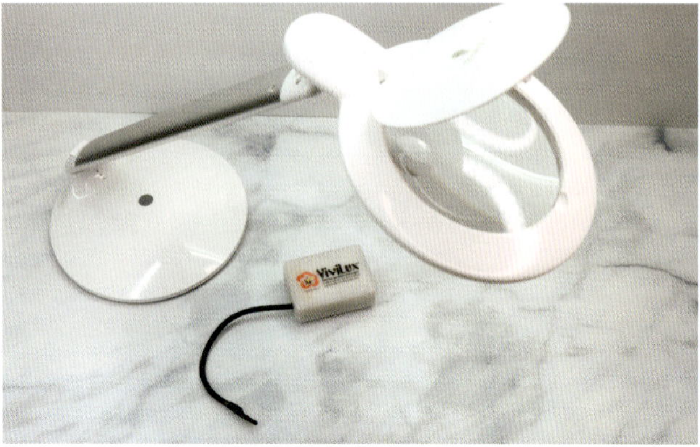

A portable light solves lighting issues in small or dark sewing spaces.

- Daylight Company—Halo Go: This portable lamp comes with a USB rechargeable battery and provides consistent brightness for a full eight hours. LED lights encircle a magnification lens for uniform lighting and 2.25X magnification over your project. An inset lens provides 4X magnification for working on the finest details. The lamp folds flat for easy portability.
- Daylight Company—UnoLamp with Clamp: This lamp allows you to see every shade and tone true to color as if in natural daylight. You can position the lamp accurately with the flexible arm and easily attach it to your work surface with the adjustable clamp. With a sleek, modern design, this lightweight lamp can be used for many tasks with its four brightness settings.
- ViviLux®—Super Bright Flexible Craft Light: The small, portable base measures just 2¼" (5.7cm) wide x 1½" (3.8cm) tall x ¾" (1.9cm) deep and houses a slim flexible light that measures 7¾" (19.7cm) long. This rechargeable light comes with both a wall plug and a USB cord and either Velcro® or magnetic backing. The 1200-lumen light is quite bright for its tiny size (compare that to a standard cell phone light of around 10 to 15 lumens!).

MACHINE-EMBROIDERY TOOLS

Machine embroidery is a category unto itself due to the added steps of hooping and placement accuracy. Here, we discuss a few of the most popular notions to help achieve perfect placement.

- dime (Designs in Machine Embroidery)—Hoop Mat: The silicone mat provides a non-skid surface, while the printed grid enables you to find the center point of any hoop by simply placing the hoop on the mat to measure. The printed area measures 22" x 16" (55.9 x 40.6cm). This hooping and measuring mat is not suitable for use as a pressing or cutting surface.
- Kimberbell®—Orange Pop Rulers: With extended corner channels, these rulers make it easy to accurately square up quilt blocks with embroidery placed on them. The open frame allows you to view your block while keeping the ruler flat and level. Each package includes no-slip grips that you can place on the backs of the rulers to keep them in place while you are cutting. They come in two sets, square and rectangle, with rulers in three different sizes. The square set also includes half-square triangle rulers in two sizes, and the rectangle set includes a bonus Flying Geese ruler, a Tumbler block template, and a standard ruler.

Machine-embroidery notions are designed to help with proper placement and accurate hooping.

- dime (Designs in Machine Embroidery)—PAL2 (Perfect Alignment Laser 2): Adjust the laser so that the outer hoop's horizontal and vertical center markings align with the illuminated crosshair. Then, after you slide the fabric under the beam and insert the inner hoop into the outer hoop, you can confirm that the fabric is square in the hoop. The laser clamps to your work surface and works with any hoop. The laser beam is protected by a reflective 6" (15.2cm)–wide and 3" (7.6cm)–tall shade. This is an ideal tool for connecting continuous embroidery designs.
- dime (Designs in Machine Embroidery)—Perfect Placement Kit: With this system, it is easy to place embroidery designs and monograms accurately on almost any kind of project. Fifteen clear plastic templates in various shapes aid with placement on linens, towels, collar corners, cuffs, and more. The set also comes with target stickers, a translucent, flexible mini ruler, and detailed instructions on using each of the templates. Also available are the Children's Perfect Placement Kit for personalizing children's items and Perfect Placement Quilt Templates.

MARKING TOOLS

Transferring marks from a pattern piece to fabric and finding the center of an area are part of the sewing process. How you mark the fabric depends on the color, material type, and even the climate where you live. In colder climates, marking-pen ink that is removed with heat can often reappear; in warmer climates, ink from air-erasing pens can take up to fourteen days to disappear. Always test your marking pen or pencil on your fabric before marking your project.

Chalk: Chalk is a traditional marking medium and a favorite of tailors. Whether in a wheel or a block, chalk is always a good option for marking because it glides on easily, is highly visible, and brushes off completely. Some chalks even wash away, which means you can leave the markings alone until you wash the project.

- Clover®—Chaco Liner Pen Style: The fine point enables you to draw accurately on both straight and curved lines. It comes in white, blue, yellow, pink, and silver, with refill cartridges available.

- Taylor Seville Originals®—Magic Chalk Liner: The smooth wheel dispenses the right amount of chalk, which either washes or brushes away. It comes in white or yellow and is refillable.
- Clover®—Triangle Tailor's Chalk: The traditional triangle shape with rounded corners marks clear and accurate lines. It comes in white, blue, yellow, and red.

Lasers: Some materials cannot be marked with a pen, pencil, chalk, or other medium, in which case a laser is an excellent option. The laser is used more as a guide, so you can use the laser light as a reference line or use stickers to keep track of your "non-markings."

- ViviLux®—3-in-1 Rechargeable Laser Light System: This portable lightweight laser can stand on its own or attach to your sewing machine with Velcro®. The system comes with alignment stickers and an alignment guide for your sewing machine. Once you find the spot you want to mark, use crosshair stickers or another marking tool. The laser is available in red or green, but the green laser is easier on the eyes.

Markers/pens: Some markers and pens are air erasing, and others are heat erasing; however, depending on the climate of where you live or where your project will live, the marks might reappear or never disappear completely. So,

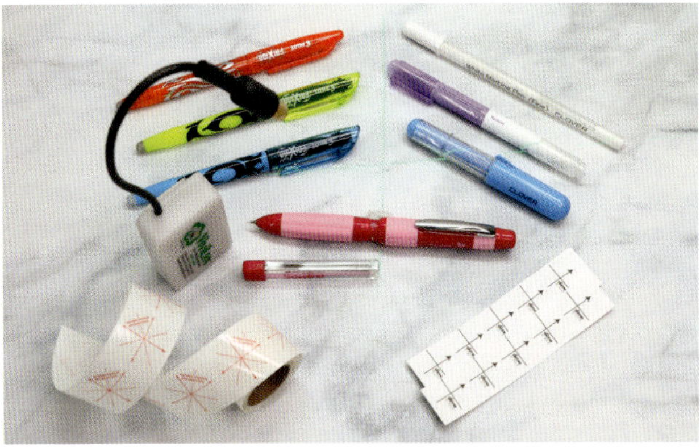

From pens to stickers to lasers, you have many options for temporarily marking your fabric.

when marking a material, it's important to test a small, hidden area first to make sure you get the desired results. It's also nice to have markers/pens in a variety of colors for working with different-colored fabrics.

- Clover®—Air Erasable Marker with Eraser: This marker has a fine felt tip, and markings will disappear on their own within fourteen days or can be removed with water.
- Pilot®—FriXion: Markings from the FriXion pens can be removed with the heat of an iron or with the friction (hence the name) of the rubber erasers at the ends of the pens.
- Clover®—White Marking Pen (Fine): This fine ballpoint pen makes white markings that disappear with the heat of an iron or when washed.

Pencils: Pencils are another option for marking fabrics; some are chalk, and others are ceramic with water-soluble dyes. As with markers and pens, test pencils on a hidden area of fabric first to see the removal results. Some pencil marks come off with water; others are removed with an eraser.

- Sewline™—Trio Colors Fabric Pencil: This pencil acts like a mechanical pencil and has three retractable lead options: white, black, and pink, with refills in a variety of colors. The lead is 0.9mm ceramic colored with water-soluble fabric dyes, and the markings can be removed with the Sewline™ Eraser.
- Clover®—Water Soluble Pencil: This pencil comes in white, pink, and blue, and the markings easily wipe off with water. To sharpen it, use a pencil sharpener just like you would with a regular pencil.

Stickers: Stickers are a nice option because they will not mark or damage the fabric. Some are clear, allowing you to see through to the fabric, while others have crosshair markings, lines, or arrows to show you where to stitch or cut.

Crosshair

- dime (Designs in Machine Embroidery)—Target Stickers: These stickers work with the PAL2 (see Lasers on page 39) for machine embroidery and quilting. Round stickers with crosshair markings make it easier to find the center. They remove easily, with no residue, and are reusable. They come in a dispenser box of 250 stickers.

- ViviLux®—Alignment Stickers: These clear, sew-through, reusable stickers work well in conjunction with the ViviLux® 3-in-1 Rechargeable Laser Light System and are ideal for quilting, garment making, bag making, and machine embroidery. Each package contains three rolls of 100 stickers each.

Ruler

- GE Quilt Designs—GEasy Ruler Stickers: Each package includes three different-colored sheets of arrow-shaped stickers (192 stickers total) that you can stick on your ruler for improved accuracy. The stickers peel off easily, leave no residue behind, and can be reused several times. You can also use them on fabric, hoops, cutting mats, and templates. They are available in two different color packs: pink/green/blue and red/yellow/purple.

MEASURING TOOLS

For your sewing projects to come out perfectly every time, you must measure for accuracy. This starts with using the correct rulers and then the proper seam allowance guides, and finishes with hem gauges.

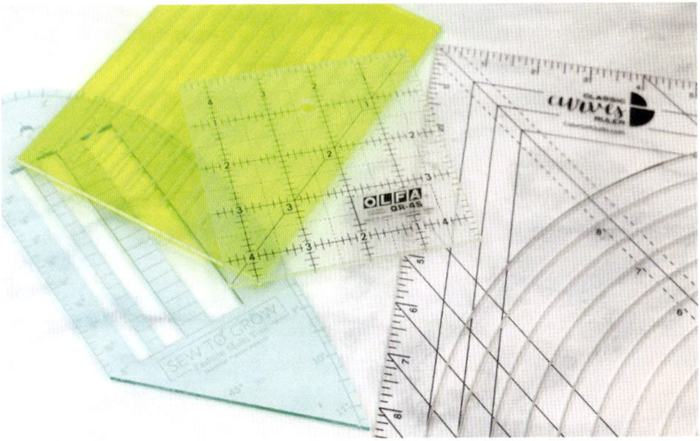

Rulers from left to right: Sew to Grow Fashion Multi Tool, Cheryl Lynch Quilts Mini Mosaic, Olfa 4" Square, and Cover Girl Quilts Classic Curves.

Rulers

Basic rulers: There are many brands of sewing, quilting, and crafting rulers on the market, but here we'll refer to the Olfa® brand. These rulers have a frosted non-glare acrylic surface, making the black imprinted lines and numbers easily visible on all fabric colors, even black. The grid lines are printed directly on clear acrylic, and the frosting is applied on top of the printing, creating a protective barrier that prevents the lines from wearing away with frequent use. The frosted surface also provides a non-slip grip for extra stability and safety when cutting. The following Olfa® rulers feature imperial measurements (inches) only.

- Long: The 6" x 24" ruler is perfect for cutting long strips, quilting, and making garments and bags. This ruler includes Olfa®'s patented compass with pivot-point angles to easily measure 15-, 30-, 45-, 60-, and 90-degree cuts.
- Short: The 6" x 12" ruler is a go-to ruler for smaller, quick cuts. Like the 6" x 24" ruler, this ruler also includes a compass with pivot-point angles for 15-, 30-, 45-, 60-, and 90-degree cuts.
- Square: Great for squaring up blocks quickly, this ruler comes in a range of square sizes from 4½" up to 16½".

Curvy rulers: Curvy rulers are designed to perform the specific task of helping you measure and cut curves with accuracy.

- Color Girl Quilts—Classic Curves Ruler: Cut ten curves from 4" to 8½" (10.2 to 21.6cm) in classic and modern patchwork shapes: Drunkard's Path, Orange Peel, Cathedral Window, Arch, Ring, and more.
- Cheryl Lynch Quilts—Curvalicious: Make rows of gently curving strips with this curved green acrylic tool. Use a 28mm rotary cutter to follow the curves closely. The tool measures 24" x 3½" (61 x 8.9mm).

Fashion rulers: These are essential tools for anyone who loves to make clothing or is serious about learning how.

- Sew to Grow—Fashion Multi Tool: You'll find many ways to use this convenient and compact tool that features a French curve, seam gauge, point turner, mitering tool, hemmer, and buttonhole placement tool. This light blue tool is made from acrylic, so you can easily mark with it and follow along its curves. It measures in imperial units. Take this tool with you to class or hang it within easy reach in your sewing room.

Slotted rulers: When following a quilting pattern, you may find that you need strips of multiple widths. Slotted rulers make cutting the various sizes out of the same block of fabric much easier. Designers also have created many patterns intended for use with slotted rulers.

- Cheryl Lynch Quilts—Mini Mosaic: This fabric cutting guide was developed to cut precise ⅜" (1cm) squares to create Cheryl's Mini Mosaic quilts; patterns are found on Cheryl's website (see Resources).
- June Tailor®—Shape Cut™: This series of slotted guides comes in various sizes to help you cut small strips with ease.
- GE Quilt Designs—Stripology Rulers: This series of slotted rulers is designed to help you easily mark strips, squares, and more. Patterns designed to use with this ruler, plus accompanying video instructions, can be found on the GE Quilt Designs website (see Resources).

Ruler accessories: From improving a ruler's grip to keeping your ruler in place to making the markings on a ruler more visible, various accessories can help you measure and cut more easily and more accurately.

Ruler accessories can help you see your ruler's markings better, keep a firm grip on your ruler, and hold your ruler securely in place.

Grippers

- Sariditty—Glitter Grippers: Glittery pink ½" (1.3cm) circles stick to the bottom of your ruler or template to prevent slipping. They come in butterfly as well as circular shapes.
- Quilter's Paradise—CutRite™ HandiGrabber: This material has a specially formulated coating to offer gripping power, and because it is extremely thin, you will not even feel it when holding your ruler or template on the fabric. It's sold in 2" x 7½" (5 x 19.1cm) pieces that you can cut into whatever sizes and shapes you need to stick on the back of your rulers and templates.
- The Gypsy Quilter®—No-Slip Grip Dots: These dots attach to the underside of your ruler to prevent slippage. The package includes both large and small dots.
- The Gypsy Quilter®—The Gypsy Gripper: This gripper attaches to smooth ruler or template surfaces with suction cups to apply even pressure to the cutting surface, allowing you to rest your hand comfortably over the handle. It also keeps your hand away from the rotary cutter, so your fingers stay safe. This tool is recommended for anyone with arthritis, carpal tunnel syndrome, or other hand issues. It comes in three sizes, so you can choose the gripper that best fits your ruler.

Magnifiers

- The Gypsy Quilter®—Spot On Dot Magnifier Lens: This tool's 3X magnification allows you to line up fabric precisely on your ruler. Stick a dot on your ruler at the spot you want to magnify. The adhesive is reusable, so with proper storage and cleaning with mild soap and water as needed, you can use the dots over and over again. The dots come in two sizes: small (1" [2.5cm] diameter) and large (2" [5cm] diameter).

Tape measures: The standard tape measure is made of vinyl and available from many different companies in a variety of colors. It is usually ½" (1.3cm) wide with inches marked on one side and centimeters on the other. The end has a metal tip, so it is easy to hold onto. Standard lengths are 60" (1.5m) and 90" (2.3m).

- Sew to Grow—Self Measuring Tape: It can be hard to measure yourself accurately for a sewing project, but this tape measure is designed to make it

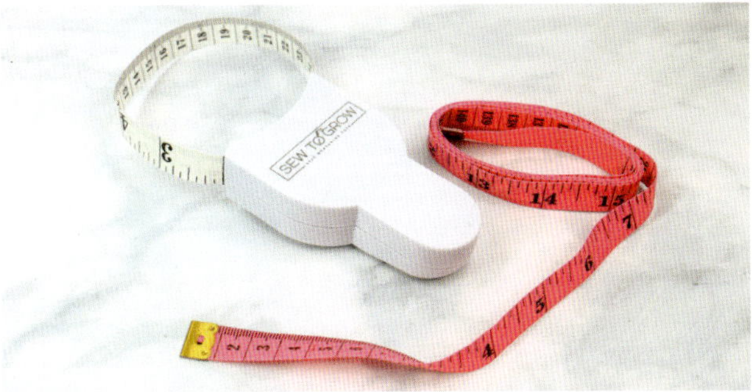

The Sew to Grow Self Measuring Tape simplifies the often-clumsy process of taking your own measurements for clothes.

easy. Simply wrap the tape measure around your body, stick the end of the tape into the opposite side of the casing, and push the button so that the tape retracts to fit the shape of your body. It's easy to look in a mirror and confirm that the tape is where you want it and level all the way around.

Seam allowance guides: The seam allowance is the distance from the stitch line to the edge of the fabric. Follow your pattern's instructions to determine the seam allowance of your project. In quilting, the most common seam allowance is ¼" (0.5cm); in bag making, it is usually ½" (1.3cm); and in garment making, it is ⅝" (1.6cm). Curved seams usually have a smaller seam allowance. Seam allowance guides help you maintain the proper seam allowance for the project you are working on.

- New Leaf Stitches—Clearly Perfect Angles: This clear vinyl guide stays in place with static, like a window cling. It is designed for creating perfect 45-degree angles with ¼" (0.5cm) and ⅝" (1.5cm) seam allowances. Simply adjust the clear sheet over the bed of your machine, sink the needle in the center, cut a hole for your feed dogs, and start stitching.
- Sew Very Smooth, LLC—Ideal Seam Gauge and Guides (Student Edition): The 4" (10.2cm) seam gauge helps you set a consistent seam allowance, the 4" (10.2cm) straight-edge seam guide helps you sew straight seams, and the 4" (10.2cm) curved seam guide aids with curved seams and features a notch that wraps around your walking foot. Both

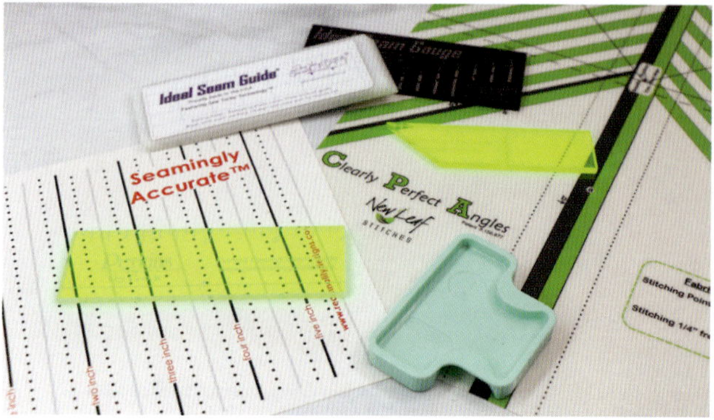

Perfect seam allowance makes for professional results.

guides feature Sew Tacky Technology™—a military-grade rubber that allows the guides to be repositioned over and over. You only need to wash it with a multi-surface cleaner, rinse it, and air-dry it to refresh its grip. Additional seam guide sizes are available.

- Perkins Dry Goods—Perfect Piecing Seam Guide: This tool positions your needle for a scant ¼" (0.5cm) seam, which is very handy when you don't have a ¼" (0.5cm) piecing foot available. You can even use it to check the accuracy of your ¼" (0.5cm) foot. To use, place the guide under the presser foot and lower the needle through the hole; you can then mark accurate seam placement on the bed of your sewing machine with tape or another marking tool.
- Clover®—Seam Guide: This 5⁄16" (8mm)–thick plastic T-shaped guide helps you sew accurately whether straight stitching, making curves, or top stitching. The handle grip makes it easy to attach and remove from your machine. The adhesive is cleanable and reusable.
- ReannaLily Designs—Seamingly Accurate™: Apply this repositionable seam guide to your sewing machine bed, over and around the metal throat plate, by cutting around the throat plate and bobbin case and then removing the cut-out portion. If you have an accessory case or free-arm, you can make additional cuts that allow you to access these

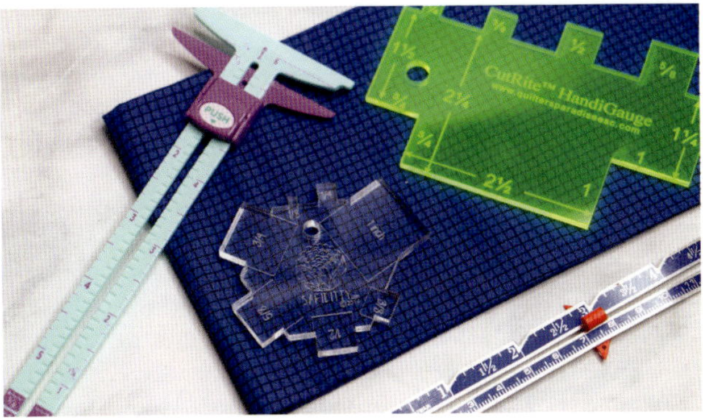

Hem gauges help you hem up, mark down, and fold over. (Fabric from Christa Quilts.)

features while not disturbing the Seamingly Accurate™. The adhesive enables you to reposition the guide as many times as needed.

Hem gauges and guides: When making a quick seam for either hand or machine sewing, these handy notions help you measure accurately.

- Clover®—5-in-1 Sliding Gauge: This handy multi-purpose tool has five uses: hem gauge, seam allowance gauge, T gauge, circle compass, and buttonhole spacing and marking tool. Comes in yellow/green and blue/purple.
- Quilter's Paradise—CutRite™ Seam Guide: Use this guide to position your machine's needle exactly where you need it for these standard seam allowances, given in imperial measurements: scant ¼", ⅜", ½", ⅝", and ¾". This green acrylic ruler also has ⅛" markings along the long side and a 45-degree end, which is useful when piecing binding strips together.
- Dritz®—Sewing Gauge: This metal hem gauge also includes a plastic slider that stays in place during use. There is a notch every inch, and metric measurements are also given.
- Sariditty—Spacer Tool: This handy spacer tool measures separation and seam spacing in the following sizes: ⅛" (0.3cm), ¼"(0.5cm), ⅜" (1cm), ½" (1.3cm), ⅝" (1.6cm), ¾" (1.9cm), and 1" (2.5cm).

TEMPLATES

This category is dedicated to the cleverly designed acrylic guides that make it easier to sew motifs and create special effects on your projects.

- Kate Colleran Designs, LLC—Braid Template: This unique template is designed to cut fabric strips at the perfect angle to achieve a braided block. Turn the block into a bag, add it to a garment, or include it in a quilt block. It comes in two sizes: the Original Braid Template and the Mini Braid Template.
- June Tailor®—Charming Circles: This circular template is 5" (12.7cm) in diameter, sized to work with 5" (12.7cm) precut squares. You can either rotary-cut a perfect circle or trace and cut it out with scissors. Start off with a circle and finish with a petal-type block.
- New Leaf Stitches—Clearly Perfect Slotted Trimmer: This tool offers an easy and precise method to square up half-square triangles. The dashed lines show you where to align the seams of unpressed half-square triangles, and then you can trim all four sides without rotating the fabric. The unique edge slots allow for simultaneous trimming of the dog ears.
- Perkins Dry Goods—Corner Cutter: This template is a fast, waste-free way to make Flying Geese, a favorite quilt block pattern. Follow the instructions to precisely cut four Flying Geese at once.
- Muppin, Inc.—Cheryl Sleboda Heirloom Smocking Template and Mini Heirloom Smocking Template: This tool simplifies the process of making Canadian smocking, an intricate-looking form of fabric manipulation, and comes in regular size and a portable mini size. The template helps you make pleating and quilting lines and more; it includes one smocking pattern. Additional patterns are found on Muppin's website (see Resources).
- Muppin, Inc.—Cheryl Sleboda Skeleton Key Pleater: This tool helps you create knife pleats and box pleats easily at your sewing machine; it works well with fabric edges and ribbon. The included instructions also show you how to create additional types of fabric manipulation.
- Quilter's Paradise—CutRite™ Slit 'n Sew Templates: Using this two-piece template set is an easier way to sew perfect quilt blocks. This

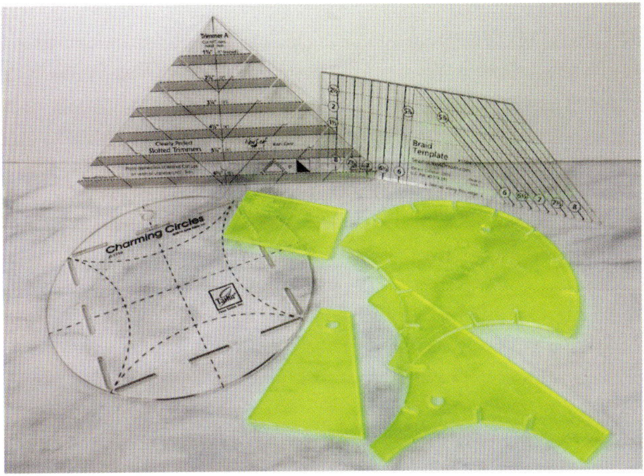

Clockwise from upper left: New Leaf Stitches Clearly Perfect Slotted Trimmer, Kate Colleran Designs Braid Template, Quilter's Paradise Clamshell Slit 'n Sew Template, Suzn Quilts Dresden Plate Template, Perkins Dry Goods Corner Cutter, and June Tailor Charming Circles.

Clockwise from upper right: Rachel Rossi Design Shorthand Template, Muppin Cheryl Sleboda Skeleton Key Pleater, Linda McGehee of Ghee's Piping Wizard, Muppin Cheryl Sleboda Heirloom Smocking Template, and Wise Craft Handmade Ruby Ruler.

brand offers template sets for many quilt block patterns, including Clamshell, Double Wedding Ring, Drunkard's Path, Melon, Tumbler, Winding Ways, and more.

- Suzn Quilts—Dresden Plate Template: These Dresden Plate block templates come in two sizes, Tiny and Mini. The Tiny makes 4½" (11.4cm) finished blocks, and the Mini makes 9" (22.8cm) finished blocks. With just a simple cut, fold, and sew method, you can make many of these petals into tiny pointed flowers.
- Linda McGehee of Ghee's®—Piping Wizard: The Piping Wizard allows you to quickly and accurately trim the seam allowance of corded piping, making it easier to add to your project, especially around curves. The manufacturer advises holding the rotary cutter on a slant to accommodate the tool's thickness.
- Wise Craft Handmade—Ruby Ruler™: This 5" (12.7cm) red square ruler is not only easy to find in your sewing room, it has a great additional function: the ruler allows you to see value differences between fabrics. It removes the "color," so only the light and dark values are shown. It comes with a free pattern.
- Rachel Rossi Design—Shorthand™ Template: This two-piece curved template set was inspired by ancient cuneiform writing. The top part of the template is a hard, durable plastic that is able to withstand rotary cuts, while the bottom is a suede-like material that sticks to your fabric, allowing you to manipulate your fabric without even lifting up the template. It comes with a free pattern.

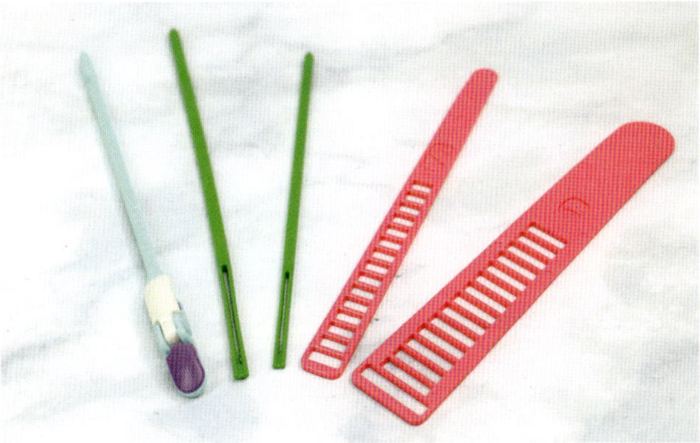

A bodkin's function is to feed small things, such as ribbon and elastic, through tubes and casing.

POINTING, TURNING, AND PUSHING TOOLS

Whether you must turn it out, feed it through, or weave it along, the items in this category are handy notions for performing these tasks.

Bodkins/threaders: This category of notions makes it easy to feed material into a tube or casing. You can even use them to weave ribbons or fabric strips for a decorative effect.

- Clover®—Clip 'n Glide Bodkin: A clip provides a firm hold on the end of the elastic, making it easy to feed the pointed end into the casing. When you pull the elastic through the other end of the casing, unclip.
- Clover®—Flex 'n Glide Bodkin: This long, flexible bodkin quickly pulls drawstring through casing. This tool works well with elastic, ribbon, and other trims.
- Wefty™—Wefty™ Needle: This flat plastic needle is used to weave fabric strips into one another. The tool has fifteen openings for fabric strips and a rounded flat end. The Wefty™ comes in a package of two, one ½" (1.3cm) and one 2" (12.7cm).

Point turners: When you sew a project with right sides together and then turn it right side out, you will need to either smooth out curved seams or

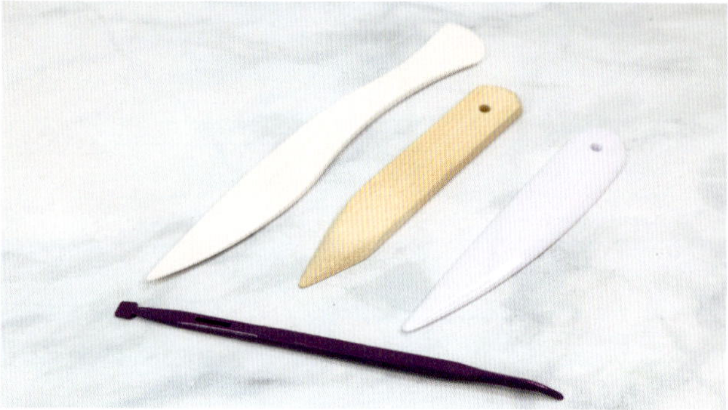

One of the handiest sewing notions, a point turner can really smooth out seams and bring crisp points to life.

give a nice, sharp finish to corners. Point turners are ideal for this function. It's important to remember that turners made from plastic should not be used with heat or steam.

- Dritz®—Quilting Point Turner & Creaser: Made of beautiful and durable bamboo, this versatile point turner and creaser comes in handy in many ways: use the pointed end to turn points in collars, flaps, belts, and appliqués or to push out corners and remove basting stitches. The creaser on the other end is used to press seams when hand- or foundation-piecing for projects. The round end also works to temporarily press and mark fabrics with a crease, while the pointed end can be used to guide fabrics when machine stitching.
- Clover®—Point 2 Point Turner: One end of this tool is curved, with a tip for point turning, and the other end is fine-tipped for detail work. It is designed to be held comfortably and securely when using either end.
- Dritz®—Point Turner and Seam Creaser: This turner is made of durable plastic and is designed to turn points in collars, lapels, and more. It's pointed on one end and rounded on the other, with a hole for a lanyard or for hanging up. To turn a point, create the seam, turn it to the opposite side, and use the pointed end of the tool to gently push out for a sharp point. To help crease a seam, use the curved end to open the

seam before pressing. For fast basting removal, clip the basting thread every few inches and use the pointed end of the tool to lift the stitches and pull them away from the fabric.

- Little Foot—That Purple Thang: There are many things you can do with this slim, 6" (15.2cm)–long purple stick. It has a curved tip on one end that works great as a stiletto, a slot that you can use as a bodkin to feed elastic through casings, and a ¼" (0.6cm) square at the other end for quick measuring.

Stilettos: A stiletto enables you to hold pieces of fabric in place and attach bindings easily. Almost like a third hand, it allows you to get close to the needle without putting your fingers in danger.

- ByAnnie—ByAnnie's Stiletto & Pressing Tool: The pressing portion of this notion is described earlier in Irons and Pressing Tools (see page 32). The stiletto helps you hold pieces of fabric together and move them as needed. It has a steel tip that is designed for durability, and the point is sand ground, which ByAnnie's website describes as being "lovingly roughed up" to prevent the point from sliding off the fabric and to help you grab, pull, and hold pieces in place.

The dual-purpose By Annie's Stiletto & Pressing Tool aids in non-heat pressing and in moving fabric pieces into place.

Tube turners: Turning a project right side out can be rewarding—but tedious. There are many ways to get good results, and some of these tools have additional helpful features, such as the ability to feed elastic into casing.

- Sue O'Very Designs—EZ Point & Turner: This unique notion is designed to easily turn projects right side out with professional results. Reimagined from a tool designed in the 1970s but discontinued in the late 1990s, this tool has multiple new features resulting from Sue O'Very's improvements on the original concept. Added features include a soft handle grip, a set screw design that allows you to open and close it easily as well as adjust the tension on the handle, a slender point for tight areas, improved ease of turning, and a cheery turquoise color. The turner can also be used as a bodkin of sorts to feed elastic, braiding, and cording inside a casing or tube. Instructional videos are available on Sue O'Very's website (see Resources).

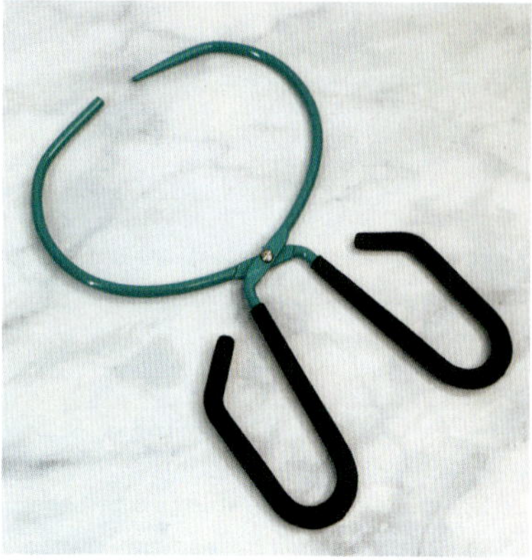

Sue O'Very Designs EZ Point & Turner is a handy tool for turning things right side out.

- Dritz®—Quick Turn Fabric Tube Turner: This handy set of cylinders and rods offers a fast way to turn fabric tubes of various sizes right side out. To use, stitch a fabric tube with right sides together, feed the appropriately sized cylinder inside the tube, and then feed about 1" (2.5cm) of the tube inside the cylinder with the corresponding rod. Keep pushing the fabric tube through the cylinder with the rod until it come out of the other end of the cylinder, right side out. This tool comes in a package of three different sizes: a 3⁄16" (0.5cm) cylinder for ⅜" to ⅝" (1 to 1.5cm) fabric tubes, a ⅜" (1cm) cylinder for ¾" to 1" (2 to 2.5cm) fabric tubes, and a ½" (1.3cm) cylinder for 1" (2.5cm) or larger fabric tubes.

QUILTING TOOLS

Quilting on a domestic sewing machine can be more challenging than on a long-arm quilting machine, but there are some wonderful tools to help make the process easier. The following specialty quilting notions are just a few of the tools that help quilters enjoy free-motion quilting on a sewing machine.

Disks/rings: Moving a baby quilt around on your sewing machine can be tricky, and a queen-size quilt even more so. Using these notions will help you maneuver the quilt with much more control by securely gripping the fabric.

- The Gypsy Quilter®—Fabulous Fabric Glide: The two squared-off hoops, one small and one large, are open on one side and designed to act as guides that control the fabric while quilting on a domestic sewing machine. These lightweight tools help you keep proper fabric tension, prevent bunching, and allow you to relax your hand position to avoid fatigue—you do not need to exert pressure on the fabric or use any other tools to grip the fabric. The textured bottom surface grips the fabric—even thick layers—securely. The guides have an inside quiltable space of 5½" (14cm) square and 7¼" (18.4cm).

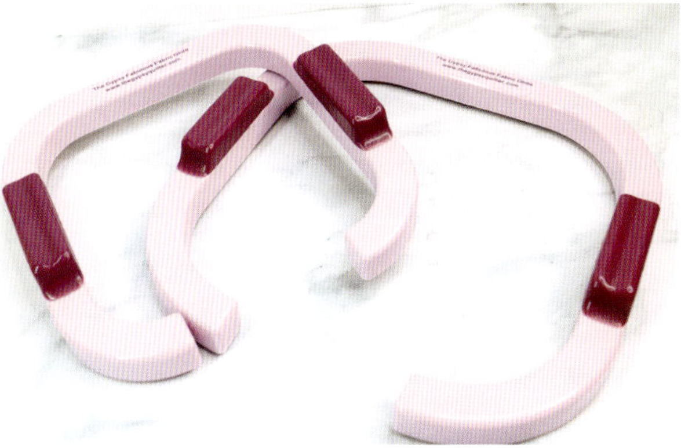

Gripping tools make free-motion quilting easy on a home sewing machine.

Quilting gloves turn your hands into grippers to easily control your project while quilting.

- Clever Craft Tools®—Grip & Stitch™ Quilting Disks: These disks come in a set of two: one large and one small. Hold the large disk in your left hand and the small disk in your right hand, place them on your quilt, and use them to move the quilt under the presser foot as you stitch. It's important to note that you should always run the machine faster than you move your hands. The disks are suitable for use with any sewing or quilting machine.

Gloves: It might seem strange to wear gloves for quilting, but these gloves transform your fingers and palms into grippers for the quilt, which makes moving the quilt around much easier.

- The Gypsy Quilter®—Hold Steady Machine Gloves: Breathable and comfortable, with gripping dots on the fingertips on one side, these gloves give you extra control where you need it. Flip the gloves over for maximum control, with gripping dots all over the palms and fingers of the other side. This versatile notion is made of 100% nylon with PVC dots; one size fits most.
- GrabARoo's®—Quilting Gloves: Made of 95% nylon and 5% spandex for a secure fit, these gloves feature tiny PVC tips on each finger to help you hold material in place. They are light and breathable for all-day wear,

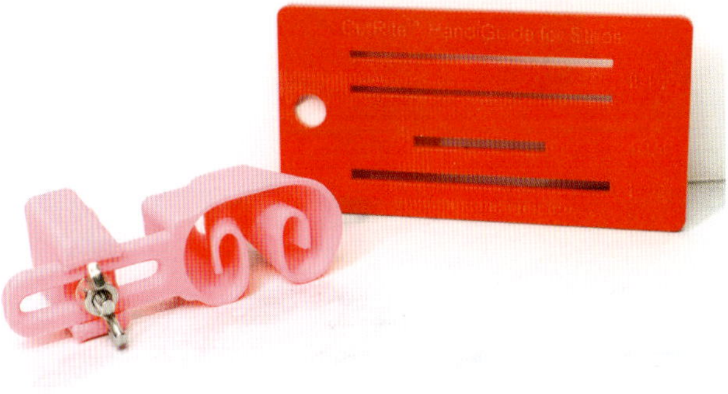

Jelly rolls are among the most popular quilting precuts, and these notions help you turn your jelly rolls into a fabric rug.

and the purple color will not transfer to your materials or machine. They come in three sizes and are machine washable.

Jelly roll guides: In the world of quilting, fabrics are sold by the yard but also in precuts, which are stacks of fabrics perfectly cut to a standard size for quilting projects. Some of the most popular precuts are charm packs (5" [12.7cm] squares), layer cakes (10" [25.4cm] squares), and jelly rolls (2½" [6.4cm] strips the width of the fabric). Precuts showcase a designer's entire fabric line, and they coordinate and work into projects nicely. A popular method of using jelly roll strips is to wrap them around long, skinny pieces of batting, connecting one to the next to form ultra-long batting-filled strips for making rugs, placemats, and baskets. Jelly roll guides help make this a simple one-step process.

- Quilter's Paradise—CutRite™ HandiGuide for Strips: This guide centers the batting strip on the fabric strip with no need for pinning. It works with 1¼" (3.2cm) or 2½" (6.4cm) batting strips in conjunction with a 2½" (6.4cm) fabric strip.
- Purple Hobbies—Jelly Roll/Batting Binding Tool: This notion helps you easily stitch together 2½" (6.4cm) fabric and batting strips for jelly roll rugs, baskets, purses, and more. It comes in a variety of colors.

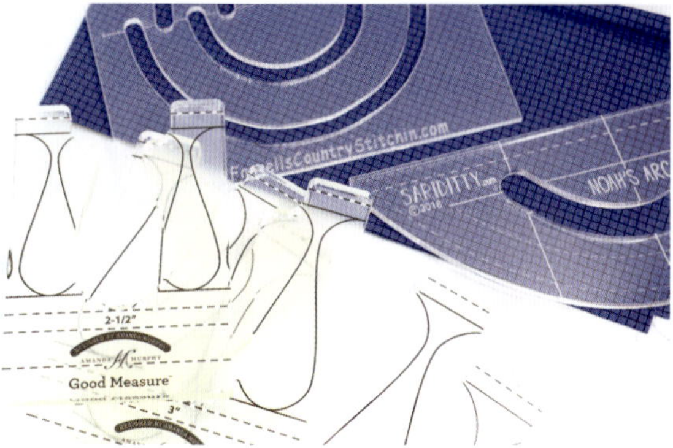

Quilting rulers feature a groove to fit your specialty quilting foot.

- Quilting rulers/templates: Quilting rulers are hard plastic rulers, sometimes referred to as quilting templates. There is a groove inside or around the outside of these rulers, allowing a specialty quilting foot to fit inside without breaking the needle or presser foot.
- Farrell's Country Stitchin—This brand offers a large variety of templates designed to work with the presser feet of various quilting machine presser feet. Their website (see Resources) has more details so you can be sure to get the right template for your machine.
- Amanda Murphy—Good Measure™: These templates are available with several different quilting motifs, each available in multiple sizes to get the perfect fit for your project. They are available for low- and high-shank machines.
- Sariditty—This unique line of quilting rulers by Sarah Thomas is designed for high-shank, low-shank, and long-arm machines.

Stencils: With quilting stencils, you can temporarily mark designs or motifs on your quilt to give you reference lines to quilt over.

- Full Line Stencil®—Stencils and Quilt Pounce™: These reusable nylon mesh stencils transfer with the use of the Quilt Pounce Pad™ and Ultimate Pounce Powder. Place the stencil over the quilt and then swipe the chalk-filled Quilt Pounce, and you're ready to add a charming motif to your quilt.

The Diamond Grid template is just one of many Full Line Stencils designed by Debby Brown Quilts.

Remove the chalk by ironing or with water. You can find stencils by dozens of designers, for example, the Diamond Grid stencil by Debby Brown Quilts (shown in the photo) at the Full Line Stencil website (see Resources).

SECURING AND ATTACHING TOOLS

Attaching trim to fabric, securing binding to the edges of quilts, and generally keeping things where you want and need them are important parts of sewing. The following notions work well for securing and attaching materials together prior to sewing them in place.

Clips

- Taylor Seville®—Magic Clips®: These clips open wide and lie flat to fit under your machine's sewing foot, so there's no need to remove them when sewing—you can sew right over them. Easy-to-read measurements are marked on the top of each clip. The clips are made with surgical-grade stainless steel and come in a variety of sizes.
- KLIPit®—Quilters Perfect KLIP®: This clip is popular with quilters because it holds fabrics very securely and is easy to work with at the same time, especially for those who have joint problems or other hand issues. With a ¾" (2cm) reach, these clips work well on corners. They also have seam guides on the bottom to simplify measuring bindings.
- Clover®—Wonder Clips®: These clips have a wide opening to securely grip layers of fabric, layered sections of projects, and heavy fabrics.

The clip base is flat, but it must be removed before the fabric passes under the presser foot. Each clip has ¼" (0.5cm) and ½" (1.3cm) seam allowance markings. They come in an assortment of colors and a variety of sizes to best suit your project.

Magnetic clips

- SewTites®—SewTites® are a handy alternative to pins and clips. This two-part magnetic system holds materials together without leaving pin holes. The top is a piece of plastic with two small yet strong magnets on the back. The bottom is a flat metal piece. Slide the bottom underneath your materials, then put the top piece in place to hold the layers together. SewTites® come in a variety of sizes to hold materials of different weights and configurations together. They work well for quilting, paper piecing, machine embroidery, and more.

Pins

- Taylor Seville®—Magic Pins™: These pins are sharp and easy to insert into materials. The comfort grip on the pin heads makes it easy to pick up and hold on to the pins. They are also heat resistant, so if you iron over them, they will not be ruined. The pins come in a storage case that snaps closed, so you don't have to worry about losing pins.

While pins have historically been used to keep materials together, clips and magnets are also very useful tools.

Organize Your Notions

Keep your sewing space tidy, and you will know where everything is and be able to get to your materials and tools quickly. Organizing can be fun, functional, and, with some of these notions, even pretty.

Bobbin holders: Bobbins are those little plastic or metal wheels wrapped with thread that fit underneath the needle area of a sewing machine. Thread evenly unwinds from the bobbin while the needle pulls the thread to the top and forms the bottom of a seam. When sewing, it is typical that your bobbin thread matches your top thread. So when your thread collection begins to expand, so does your bobbin collection, and, therefore, you need a way to organize your bobbins. When choosing a system for organizing your bobbins, make sure you choose holders that are correctly sized for your type of bobbins.

- Purple Hobbies—Bobbin Clips: These clips help you store and organize your bobbins with no more tangles. Each plastic clip holds three metal or plastic Style A (Class 15) bobbins (standard size for most home machines), and you just snap them in. The clips also make it easier to bring your bobbins with you to class or wherever you are working on your project. Bobbin Clips are available in a variety of colors.
- SCHMETZ—BobbinSaver®: With its hollowed-out circular shape and flexible material, this bobbin holder grips plastic and metal bobbins securely, keeping your bobbins organized and your thread tangle-free. This helpful notion comes in three sizes: regular, which holds approximately twenty standard-size (Class 15) bobbins; Class M, which holds the larger bobbins used in many commercial machines; and jumbo, which holds Series 7 and 8 BERNINA bobbins as well as some others used in industrial machines.

The SCHMETZ BobbinSaver (left), Purple Hobbies Bobbin Clips (top right), and Gypsy Quilter Tulips Bobbin Clamps (bottom right).

- The Gypsy Quilter®—Tulips Bobbin Clamps: Each rubber clamp securely holds one bobbin. Thread the clamps onto the included chain to keep them together or to bring them with you. The clamps work with Style A (Class 15) and L bobbins. These brightly colored notions come in packages of twelve or sixty.

Fabric organizers

- The Fabric Organizers: Easily fold and store your fabric stash with these organizers. Made of a stiff, acid-free material, each of the three sizes (large, small, and shorty) has a rounded rectangle shape with tabs to hold fabric in place. You can store your organizers—which function like mini-bolts—side by side, stacked, or in bins so that the fabrics are easily visible and you can find what you are looking for.

The Fabric Organizers function like mini fabric bolts that streamline your fabric collection. (Fabric from Christa Quilts.)

Needle storage: Because of their small, thin size, hand-sewing needles are easy to lose track of. Using a storage device for needles is a clever way of keeping track of them. In some cases, you can even store your threaded needles, which makes it easy to keep a few needles threaded and ready for stitching.

Machine needles are designed to be used for six to eight hours and then discarded. If you've used a machine needle for only a short time, you do not need to discard it, but you should not put a used needle back into a case of new needles. It's ideal to store your needles in a way that allows you to identify their type and size as well as keep new and used needles separate.

- Clover®—Dome Threaded Needle Case: This is a unique, tangle-free storage option for up to ten threaded hand-sewing needles shorter than 2" (5cm) with thread tails of 35" (80cm) or less. The case uses a grooved reel—place a needle into one of the grooves, insert the thread into the slot, and then turn the reel to wind. To remove the needle, hold the needle and thread simultaneously, and gently pull it out. There is a clear cap that screws onto the top of the case.

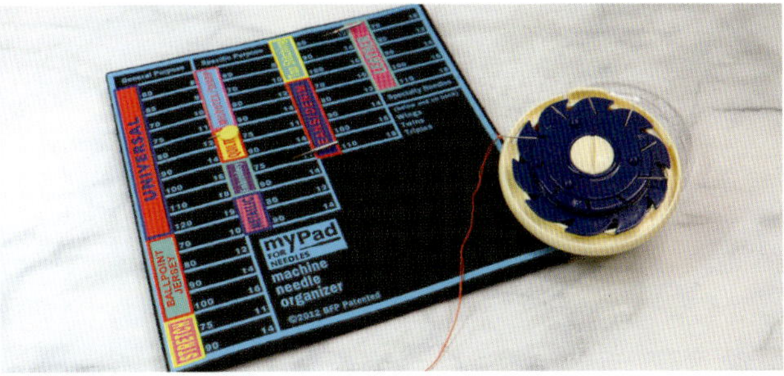

The SCHMETZ myPad (left) offers labeled storage areas for sewing machine needles, while the Clover Dome Threaded Needle Case offers tangle-free storage for threaded hand-sewing needles.

- SCHMETZ—myPad™ for Needles: If you are frustrated by unidentified sewing machine needles, you'll appreciate this 6.¾" (17.1cm) square felt pad with different-colored areas labeled by needle size and type. Just stick the gently used needle into the correctly labeled area of the thick felt pad, and it will be ready for your next project. The pad also provides spaces that you can label yourself for unusual sizes or specialty needles.
- Dritz®—Needle Storage Tubes: Designed for hand needles, these clear plastic tubes hold needles up to 2¼" (5.7cm) in length, and one tube can hold twenty needles. When you open the tube, a magnet holds the needles securely and fans them out for selection. Each package includes three tubes with different-colored caps and thirty-three pre-printed adhesive labels to designate the size and type of your needles.

Pin/clip storage: Pins and clips hold your fabric together for either cutting or sewing. Storing them in a container keeps them in one place and easy to move around your sewing space. Because pins are typically made of metal, many pin organizers use magnets to keep your pins safely organized.

- Magnetic pin cushions
  - SCHMETZ—Grabbit® Magnetic Pin Cushion: With its easy-to-hold round design and shallow, concave top surface, Grabbit makes picking up pins easy. It measures 1 5/16" (3.3cm) tall, with a 4¼" (10.8cm) diameter. Its powerful magnet sweeps up pins from your work surface and keeps them secure in an organized pile. It includes

Different types of magnetic pin cushions, clockwise from left: the Zirkel Magnetic Organizer, Purple Hobbies Magnetic Pin Cup, and SCHMETZ Grabbit.

fifty color-ball plastic-head steel pins and comes in a range of colors. Grabbit's magnet is safe to use around computerized sewing machines.

- SCHMETZ—Wrist Grabbit® Magnetic Pin Cushion: The Wrist Grabbit features the same magnetic power of the regular Grabbit, but in a handy wristband with a magnet measuring 2" x 1½" (5 x 3.8cm). Move your wrist over your work surface to sweep up pins, and they will be held, heads aligned and ready to grab, on the wristband—practically at your fingertips. The package contains seventeen color-ball plastic-head steel pins.
- Purple Hobbies—Magnetic Pin Cups: These handy cups hold your pins point down for safety. Each cup comes with a lidded case for safe storage and transport. The cups come in two sizes: standard, for pins up to 1½" (3.8cm) long, and large, for pins up to 2" (5cm) long. They come in a variety of colors.
- Zirkel®—Zirkel® Magnetic Organizer: Drop a handful of pins onto this innovative notion's square, magnetic surface, and they will immediately be arranged into a flat, circular sunburst pattern, spread out and ready to grab. It works with any item attracted by magnets: pins, metal clips, screws, and more.

- Clip holders
  - KLIPit®—KLIPit Kubby: This unique stackable clip storage works with the Quilters Perfect KLIP™ clips. It holds forty clips, and the trays rotate on a lazy Susan base with even more storage in the center holder. It's easy to move and take with you.
  - Clover®—Wonder Clips Wrist Cushion: This comfortable, wearable holder made from felted petals attaches to a snap wristband and holds up to twenty-five regular or fifteen jumbo Wonder Clips®. The package also includes three regular-sized clips. Attach each clip to two or three petals at a time for a secure grip. The wristband fits most wrists, and it's easy to remove and reattach the wristband.

Designed to hold Clover's Wonder Clips, the brand's wearable Wrist Cushion keeps clips close at hand.

Scissor case: A scissor case not only stores your scissors safely, it also protects your blades to keep them sharp and in top cutting condition.

- ScissorFobz®—Scissor Sheaths: These sheaths come in a variety of sizes to cover your scissors' blades and protect them. The proprietary ScissorGripper™ material inside the sheaths keeps the scissors from slipping out. With a variety of outer materials, designs, and colors, they

A ScissorFobz® Scissor Sheath protects a pair of Famore Cutlery's Pro Cut Comfort Handle Fabric Shears.

come in a pack of four sizes: 3½" x 1" (8.9 x 2.5cm), 4¼" x 1¼" (10.8 x 3.2cm), 5½" x 1¾" (14 x 4.4cm) and 6" x 2" (15.2 x 5cm).

Thread holders/stands: When thrown in a drawer, thread tends to get tangled, knotted, or otherwise messy. Keeping your thread collection nice and tidy is important to having an organized sewing space. Your sewing machine's built-in stand is not always the best thread-delivery option, depending on the type of thread or application. A separate thread stand keeps thread from twisting or jamming up your machine.

- The Gypsy Quilter®— Buddies Bobbin Clips: These handy clips keep your bobbins paired up with the matching thread. Just squeeze the ends together and insert through the middle of the bobbin and the matching thread spool. They fit most sizes of bobbins and spools and come in a jar of 120 in multiple colors.
- The Gypsy Quilter®—Peels Spool Huggers: Keep your thread from unraveling by wrapping one of these brightly colored rubber "huggers" around each spool. The flexible rubber will fit spools of any size.
- Superior Threads®—Superior Thread Holder: This is an innovative thread-delivery tool that works with spools and cones up to 3½" (8.9cm) in diameter. It includes threading options for specialty threads and can be moved to different positions relative to your sewing machine, depending on the kind of thread you are using. This thread stand helps reduce twisting and other problems that can occur with thread.
- Superior Threads®—Telescoping Metal Thread Stand: This sturdy, portable thread stand is adjustable in height from 6" to 15" (15.2 to 38.1cm) and makes it easier to work with specialty threads and large-diameter spools.

The Gypsy Quilter's Peels Spool Huggers fit around spools of all sizes to keep thread from unraveling, while the brand's Buddies Bobbin Clips pair matching spools and bobbins.

- Quilter's Paradise®—The Thread Stem: This tool allows you to use larger spools and cones that your home sewing machine's spool pin cannot accommodate. Attach it to your machine by inserting the spool pin through the Thread Stem's larger hole and then putting the thread through the smaller hole at the other end. It works with almost all types of machines, and its small size makes it easily portable.

Tabletop organizers: When you are in the middle of a sewing project, you want your sewing tools and notions organized but still handy. These organizers keep your workspace neat and tidy, and some are portable so you can keep your tools together as you move around your sewing space.

- Doohikey Designs—Ruler Roundup™: This sturdy wood organizer is available in two sizes for domestic and long-arm rulers and in four bright colors. There are seven slots ready to hold rulers, patterns, templates, cutting mats, and small notions in between the rulers. The back is flat and the front is curved, making it a perfect fit up against a wall or window.
- Purple Hobbies—Phone and Notion Caddy: This notion holds your cell phone, glasses, scissors, marking pens, rulers, and more. The phone

Tabletop organizers clockwise from upper left: the Doohikey Designs Ruler Roundup, Purple Hobbies Phone and Notion Caddy, Hummingbird Highway Your Nest, and It's Sew Emma Stash 'n Store.

stand allows you to set up your phone in portrait or landscape mode for watching videos or charging. It's available in a variety of colors.

- The Gypsy Quilter®—Sensational Sip & Snip: This handy cupholder and accessory bag clamps to your work surface to keep essentials nearby without taking up space. Use the removable bag for sewing tools, notions, and other small items, or sweep scraps and clippings into it. The rubberized clamp will not damage your work surface. The cupholder is made of rust-proof coated metal.
- It's Sew Emma®—Stash 'n Store: This tabletop caddy's durable silicone pockets hold your tools in an upright position, allowing you to find and grab them quickly. The flexible silicone provides a firm grip and is easy to clean. It's available in two sizes—the large size measures 1½" x 2" x 8¼" (3.8 x 5 x 21cm), and the mini size measures 1½ x 2" x 4" (3.8 x 5 x 10.2cm)—and a variety of colors: lime, pink, gray, white, purple, red, mint, gray, navy, and aqua.
- Hummingbird Highway—Your Nest: With an approximately 4½" (11.4cm) base, and 1½"(3.8cm) tall, this sturdy, flexible organizer holds sewing tools and notions, your cell phone, patterns and other papers, and small, easy-to-lose items. It's available in seven avian-inspired colors, such as Cardinal, Peacock, and Raven.

Other sewing-room tools

- Sitting surfaces
  - The Gypsy Quilter®—Sit Upon: Sit on this durable, rubberized inflated disc anywhere you sew or quilt. It works on any type of chair, and you can even put it on your car seat for added comfort when driving or on your office chair for extended periods of work. Much like an exercise ball, it helps strengthen your core, correct your posture, and improve circulation, all the while relieving lower back pain by putting less pressure on your tailbone and ultimately reducing fatigue. The nubby texture allows air flow to keep you cool, but you can also place the nubby side on the chair for a smoother seating option.
- Cleaning tools: Keeping your sewing machine clean is important for it to run smoothly. Professionals recommend having your machine serviced every one to two years, with regular cleaning and maintenance

Clean your sewing machine at home, in between professional servicing, to keep it running smoothly.

at home. The notions discussed here will help keep your machine free of lint and bits of leftover thread.

- Machine lint brushes: Most sewing machines come with their own lint brushes; if yours did not, you can buy one online or at almost any sewing shop. This small brush is designed to clean inside your sewing machine and remove lint buildup.
- Micro vacuum attachments: You can buy this type of attachment for your regular vacuum cleaner at a sewing shop or online. It converts your vacuum into a powerful tool that can clean the smallest and hardest-to-reach areas. It easily attaches to any vacuum hose to clean sewing machines, computers, office machines, auto interiors, and stereo and video equipment.
- It's Sew Emma®—Oh Sew Clean Brush and Cloth Set: This set of soft-bristled flat and angled brushes plus a microfiber cloth will help you clean every corner, crevice, and surface of your sewing machine. It's available in gray, teal, and pink.

Resources

In addition to the author's website (Sue O'Very Designs: *www.SueOVeryDesigns.com*), check out the following sources of handy notions, information, and inspiration.

Amanda Murphy: *www.amandamurphydesign.com*
Angela Wolf Patterns: *www.angleawolfpatterns.com*
Attached, Inc.: *www.mistyfuse.com*
Bear Thread Designs, Inc.: *www.bearthreaddesigns.com*
Bo-Nash: *www.bonashstore.com*
ByAnnie: *www.byannie.com*
Cheryl Lynch Quilts: *www.cheryllynchquilts.com*
Christa Quilts Fabric: *www.christaquilts.com*
Clever Craft Tools: *www.clevercrafttools.com*
Clover: *www.clover-usa.com*
Color Girl Quilts: *www.colorgirlquilts.com*
Cricut: *www.cricut.com*
Daylight Company: *www.daylightcompany.com*
Debby Brown Quilts: *www.debbybrownquilts.com*
dime (Designs in Machine Embroidery): *www.shop.dzgns.com*
DIYStyle: *www.diystyleshop.com*
Doohikey Designs: *www.doohikeydesigns.com*
Dritz: *www.dritz.com*
The Fabric Organizer: *www.thefabricorganizer.com*
Famore Cutlery: *www.famorecutlery.com*
Farrell's Country Stitchin: *www.farrellscountrystitchin.com*
Faultless: *www.faultless.com*
Full Line Stencil: *www.fulllinestencil.com*
GE Quilt Designs: *www.gequiltdesigns.com*
Ghee's: *www.ghees.com*
GrabARoo's: *www.grabaroos.com*
The Gypsy Quilter: *www.gypsyquilter.com*
Hummingbird Highway: *www.hummingbird-highway.com*
It's Sew Emma: *www.itssewemma.com*
June Tailor: *www.junetailor.com*
Kate Colleran Designs, LLC: *www.seamslikeadream.com*
Kimberbell: *www.kimberbell.com*

KLIPit: *www.klipit.us*
Little Foot: *www.littlefoot.net*
Mary Ellen: *www.maryellenproducts.com*
Muppin, Inc.: *www.muppin.com*
New Leaf Stitches: *www.newleafstitches.com*
Odif: *www.odifusa.com*
OESD: *www.embroideryonline.com*
Olfa: *www.olfa.com*
Oliso: *www.oliso.com*
Quilter's Paradise: *www.quiltersparadiseesc.com*
Panasonic: *www.shop.panasonic.com*
Perkins Dry Goods: *www.perkinsdrygoods.com*
Pilot: *www.pilotpen.us*
Purple Hobbies: *www.purplehobbies.com*
Rachel Rossi Design: *www.rachelrossi.design*
ReannaLily Designs: *www.reannalilydesigns.com*
Sariditty: *www.sariditty.com*
SCHMETZ: *www.schmetzneedles.com*
Scissorfobz: *www.scissorfobz.com*
Sew Fine Thread: *www.sewfine.ca*
Sewline: *www.sewline.com.au*
SewTites: *www.sewtites.com*
Sew to Grow: *www.sewtogrow.com.au*
Sew Very Smooth: *www.sewverysmooth.com*
Soak: *www.us.soakwash.com*
Sulky: *www.sulky.com*
Sullivans: *www.sullivansusa.net*
Superior Threads: *www.superiorthreads.com*
Suzn Quilts: *www.suznquilts.com*
Taylor Seville: *www.taylorseville.com*
TQM Products: *www.tqmrulers.com*
Traditional Primitives: *www.traditionalprimitives.com*
ViviLux: *www.viviluxusa.com*
Wefty: *www.weftyneedle.com*
Whole Circle Studio: *www.wholecirclestudio.com*
Wise Craft Handmade: *www.wisecrafthandmade.com*
Zirkel: *www.thezirkel.com*

About the Author

Sue O'Very-Pruitt is known for her fresh style and love for all things sewing. Whether on her YouTube channel for machine embroidery (*SueOVeryTV.com*), on her channel dedicated to sewing and sewing-notion reviews (*SookieSewsTV.com*), or when teaching in person, Sue has a knack for simplifying sewing and making the many details memorable and easy to understand. Drawing on her degree in fashion design, experience in leather work, and years of designing and making costumes, she enjoys teaching sewists of all experience levels how to sew, how to use sewing notions, and how to machine embroider. With the love and support of her husband, Joey, and her daughter, Heidi, Sue continues to create new and clever ways to bring all things sewing to her audience. You can find Sue on Instagram, Pinterest, and Facebook *@SueOVeryDesigns*, on her Sookie Sews social spots *@SookieSews* on Instagram and Facebook, and on her websites *www.SueOVeryDesigns.com*.and *www.SookieSews.com*.